Minerals Processing and Beneficiation

About The Centre

The Centre for Science and Technology of the Non-Aligned and Other Developing Countries (NAM S&T Centre) is an inter-governmental organisation with a membership of 47 countries spread over Asia, Africa, Middle East and Latin America. Besides this, 12 S&T agencies and academic / research institutions of Bolivia, Botswana, Brazil, India, Nigeria and Turkey are the members of the S&T-Industry Network of the Centre. The Centre was set up in 1989 to promote South-South cooperation through mutually beneficial partnerships among scientists and technologists and scientific organisations in developing countries. It implements a variety of programmes including international workshops, meetings, roundtables, training courses and collaborative projects and brings out scientific publications, including a quarterly Newsletter. It is also implementing 5 Fellowship schemes, namely, NAM S&T Centre Research Fellowship, Joint NAM S&T Centre – ICCBS Karachi Fellowship, Joint CSIR / CFTRI (Diamond Jubilee) - NAM S&T Centre Fellowship, Joint NAM S&T Centre – ZMT Bremen Fellowship and Research Training Fellowship for Developing Country Scientists (RTF-DCS) in Indian institutions. These activities provide, among others, the opportunity for scientist-to-scientist contact and interaction, training and expert assistance, familiarising the scientific community on the latest developments and techniques in the subject areas, and identification of technologies for transfer between member countries. The Centre has so far brought out 63 publications and has organised 94 international workshops and training programmes.

For further details, please visit www.namstct.org or write to the Director, NAM S&T Centre, Core 6A, 2nd Floor, India Habitat Centre, Lodhi Road, New Delhi-110003, India (Phone: +91-11-24645134/ 24644974; Fax: +91-11-24644973; E-mail: namstcentre@gmail.com; namstct@bol.net.in).

Minerals Processing and Beneficiation

— *Editors* —
Awni Al-Otoom
Mohammad Al-Harahsheh

CENTRE FOR SCIENCE & TECHNOLOGY OF THE NON-ALIGNED AND OTHER DEVELOPING COUNTRIES (NAM S&T CENTRE)

2015
DAYA PUBLISHING HOUSE®
A Division of
ASTRAL INTERNATIONAL PVT. LTD.
New Delhi – 110 002

ISBN 9789351305484

Published by : **Daya Publishing House®**
A Division of
Astral International Pvt. Ltd.
– ISO 9001:2008 Certified Company –
4760-61/23, Ansari Road, Darya Ganj
New Delhi-110 002
Ph. 011-43549197, 23278134
E-mail: info@astralint.com
Website: www.astralint.com

Laser Typesetting : **Classic Computer Services**, Delhi - 110 035

Printed at : **Replika Press Pvt. Ltd.**

PRINTED IN INDIA

Foreword

It is widely reported that the number of resource driven countries is continuing to increase to an extent that some estimate it to have increased by more than 40 per cent since 1995. However, this increase has not been accompanied by high activity in the industrialization in these countries. In most of these countries this is due to a lack of further investment in the development of resources that could trigger economic and social transformation. Some of the reasons for not developing these resources are historical, in that most of the resource rich countries were in the past been used as exporters of raw material without using opportunities of developing these resources into end products that could further benefit the citizens of the host country. All these benefits were enjoyed by the importers who happened to be first world countries. In so doing all the expertise of mineral processing and beneficiation were established outside these resource rich countries.

Availability of natural resources and resource-based industries is essential to prosperity and well-being of a country in this rapidly changing world. Mineral resource is an important value-creating asset for the developing countries and mining and consumption of minerals is indispensable for them to improve the quality of life. It is very critical that developing countries continue to develop the mineral resources into products that can contribute to positive economic growth, development of necessary skills.

I am encouraged by the rebirth of true patriotism where a number of resource driven countries are now forcing resource operators to do more if not all mineral processing and beneficiation business opportunities in the resource hosting jurisdictions. This if done where these countries have a competitive advantage, it will boost domestic capabilities, technology opportunities and skills development in the host countries. At the end, this will in turn convert natural resources wealth into long term prosperity for the resource rich nations. This can only be achieved if the the policy makers, scientists, technologists, academics and other industry experts come together and assess ways and means to address the aforementioned skills gap.

I am therefore delighted to learn that during 25-27 September 2012, the Centre for Science and Technology of the Non-Aligned and Other Developing Countries (NAM S&T Centre) jointly with the Department of Science and Technology of South Africa organised the International Workshop on Minerals Processing and Beneficiation in Johannesburg, South Africa. I understand that the event was successful in highlighting the importance of the issues related to Mineral Sector to unravel its wide range of possibilities so as to provide a more sustainable foundation for the developing world's transition from a resource enriched condition to an industrial economy. The present book titled "Minerals Processing and Beneficiation" is an outcome of these deliberations and comprises 17 scientific and technical papers from the experts and professionals of 16 countries.

I compliment the NAM S&T Centre for bringing out this highly valuable publication and am confident that it is a relevant step forward that will help the developing countries in working out appropriate strategy, policies and programmes on the exploitation of new technologies and developing trained human resources on mineral processing and beneficiation for a better industrial economy.

Mr. Abiel Mngomezulu,
President and CEO,
MINTEK,
Randburg, South Africa

Preface

Minerals exist in abundance throughout the world, particularly in most countries of the Non- Aligned Movement group (NAM). It is believed that minerals in these countries could alleviate their economical burdens and poverties. However, the current utilization of minerals is considered at its minimal in these countries. Different challenges are causing the underutilization of these minerals. Many of these challenges are due to the lack of local expertise, the absence of the appropriate policies, and many other challenges.

It is, therefore, imperative for developing countries to take appropriate policy measures for building the capabilities of local expertise, acquire and adapt new technologies; enhance efficiency of production and reduce production costs; and undertake total quality control, quality assurance and good manufacturing practices.

This book is an extended and updated version of the papers presented at the International Workshop on 'Minerals Processing and Beneficiation' which was organized in Johannesburg, South Africa during 25-27 September, 2012 by the Centre for Science and Technology of the Non-Aligned and Other Developing Countries (NAM S&T Centre) jointly with the Department of Science and Technology (DST), South Africa. The workshop brought together various stakeholders, *viz.* experts, policy makers, researchers, technologists and entrepreneurs to share knowledge, approaches and experiences. The contents of this book is divided into two parts. **Section 1: 'Trade and Technology'; and Section 2: 'Policies and Strategies'.**

Section 1 includes thirteen papers, which provide general perspectives on the current status of mineral processing and technologies in different NAM countries.

The Status of the Gambia's Mining Industry is presented by Alieu Jawo. The government of Gambia is committed to ensuring that mineral exploitation benefits the population. An example of mineral utilization is presented through which the sand is utilized in a two-step process before export. This consists of a primary and a secondary separation stages. In this way, opportunities are created for the employment of Gambians along the value chain. Government emphasizes linkages - forward, backward and lateral.

Akufo-Kumi *et al.*, have employed X-ray Diffraction-Energy-dispersive X-ray spectroscopy (XRD-EDS) for the analysis of Ekon Kaolin in Ghana. It was found that Ekon Kaolin have high purity with properties that make it a potential raw material for the production of alumino-silicate refractories suitable for non-ferrous industries.

Dianne McDonald has presented a status update of an initiative by the Guyana Geology and Mines Commission to Improve Mineral Recovery, and increase profitability in the local mining sectors. It was reported in this paper that small and medium scale operators in Guyana have improved their productive capacity over the years. They have moved from the simple pick axe and shovel to gravel pumps and excavators which allow them to move tonnes of material per day. Additional initiatives include improving current practices; promoting improved recovery methods by demonstrations; promoting mercury free methods for gold extraction; educational and awareness campaigns by seminars, brochures, flyers; technical assistance campaigns to mine operators.

Agustinus *et al.* presented a paper on the utilization of Indonesian Java natural silicates mineral and recycled amorphous silica and its beneficiation for the production of advanced materials. Perlite, bentonite and zeolite are enormous in large areas surrounding the coastal line in Tasikmalaya- West Java. Perlite and bentonite have been used as a raw advanced material for preserving the microorganism/bacteria called *Bacillus licheniformis* (*L.P*) and used for reducing the phenols and other organic compounds concentration in waste water treatment. The products are formed in preserved microorganism tablet.

Al-Otoom has presented a paper on the characterization and beneficiation of Jordanian oil shale. The beneficiation utilizes froth floatation and demineralization using different acids. The cauterization of Jordanian oil shale was performed using few techniques include scanning electron microscope with energy dispersive spectrum analysis, X-ray diffraction, and Qemscan. The dominant mineral phase found in the oil shale was the calcite. Different proportions of clays, quartz, and phosphate were also present. On the beneficiation side, froth floatation of Jordanian oil shale did not show any significant improvement in the shale oil yield. However, demineralization is considered a technically viable process for the beneficiation of oil shale.

Githinji *et al.*, have presented an overview of mineral processing and beneficiation in Kenya. Mining activities in Kenya are currently dominated by production of non-metallic minerals encompassing minerals such as soda ash, fluorspar, kaolin and some gemstones. Although faced by many challenges, the vast endowment of minerals in Kenya shows that the mining industry has a great potential to contribute more than 15 per cent of the national GDP. Authors believe that this projection can only be achieved if the miners, technologists and researchers worked on how to overcome the challenges and find better ways of enhancing mineral processing and beneficiation techniques.

Ismail Ibrahim and Md. Muzayin Alimon have presented a paper to demonstrate the results obtained from a bench scale feldspar flotation for Malaysian pegmatite ore. The bench scale utilizes non hydrofluoric modifier and mixed collector. The behavior of duomeen TDO in the feldspar–quartz flotation system has been investigated through zeta potential and Fourier Transform Infrared (FTIR) measurements. Several series of feldspar flotation tests were carried out using different set of flotation parameters. The results show that high feldspar grade concentrate (86.79 per cent) and a good recovery (87.96 per cent) can be obtained at pH 2 by using 800 g/t duomeen TDO and 90 g/t Aero 65.

Wuyep *et al.* have presented the main challenges facing the mining of columbite in Nigeria. The main challenge reported is the lack of information on the actual amounts of the current reserve of columbite. In addition, the lack of modern and innovative research, and the lack of skills and modern equipment is considered one of the main challenges reported by the author.

Nwokemodo *et al.* have also presented the challenges facing the mineral processing and beneficiation of Nigerian brine. Authors have also presented an improved method in the processing of naturally occurring brine in Nigeria through which production increased to 56kg per 8 hours from 12-15 kg in 28 days. In addition, the new improved method has reduced the amounts of impurities making it usable by humans.

Komba and Kayandabila have presented the Tanzanian prospective on the mineral beneficiation and value addition. Tanzania went through major economic reforms commencing in the late 1980's that have had an impact to the minor reforms in the mining sector today. As a result of use of good macro-economic policies set in place, the first Mineral Policy of 1997 and enactment of the Mining Act of 1998 was in place. This process of policy formulation and mining code implementation was a crucial milestone in the country's mineral sector. After many years of implementing the mining Act, priority is given to mineral beneficiation and value addition processes.

Balu Tabaaro presented a paper on Uganda's mineral processing and the challenges facing this industry. The author believes that mineral concentrates are exported outside Uganda at low cost due to the lack of local expertise in the beneficiation and purification of minerals. Therefore, the author suggests that Non-Aligned Movement (NAM) countries need to invest in training of scientists and technologists in the mineral processing of their mineral resources. They also need to train in skills for upgrading and adding value to these resources so that products can be made locally and jobs can be created.

Lungu presented a paper on the development and status of Zambian manufacture of industry spares from copper based alloys. The author refers to a report that noted that with the copper industry output peaking at 1.5 million tonnes per year and the annual revenue would increase to US$4.0 billion. The above report elaborated that Zambia's copper mines' cost of production was considered high due to the high cost of equipment and spares and other consumables. In addition, the report refers to improved linkages between the mines and local companies or suppliers can impact greatly on job creation as well as improving the quality of the jobs.

Sehgal has presented the technology of bioleaching in mineral processing and beneficiation. Bioleaching technology can prove, according to the author, to be very promising technology particularly in the developing countries. The most commonly used microorganisms for bioleaching technology are *Thiobacillus thiooxidans* and *Thiobacillus ferrooxidans*, which are chemolithotrophic bacteria converting the insoluble metal sulphides into soluble metal sulphates. They usually act either by coming in contact with the ores containing minerals and causing oxidation of minerals or by producing leaching agents which oxidize the metals.

Section 2 includes four papers and presents Policies and Strategies of mineral processing in some of the NAM countries and the potential improvements.

Yehia presented a paper on the current status of Egyptian mineral industries. The Egyptian mineral industry has enormous growth potential for the next decade owing to the country's abundant mineral resources- especially gold, natural gas, phosphate rock, and tantalum and the increased demand for steel reinforcing bar and cement by the construction sector. The author believes that Egypt possesses many investment attraction factors. These factors include; a strong infrastructure of transportation and communication, cheap energy sources, skilled manpower, eligible industrial cities, a strong and secure banking system, and an active stock market.

Kishore Boodhoo and Roshan T. Ramessur have presented a paper on the potential exploitation of deep sea minerals in Mauritius. The deep sea minerals include manganese nodules and marine polymetallic deposits. Explorations are currently being conducted along segments of the Central Indian Ridge in the Indian Ocean by Korean and German teams in search for metalliferrous deposits. The authors believe that there is a need to develop a viable and sustainable marine mineral industry in Mauritius and to develop the sustainable management of deep sea mineral resources with improved human, technical and effective monitoring systems for exploration of different potential areas in Mauritius.

Tran Thanh Hai has presented an overview of the potential mineral resources and current mining practices in Vietnam. The mineral resources comprise many types of metallic, non-metallic and industrial minerals with thousands of deposits and occurrences have been discovered. Metallic resources consist of ferrous, base, precious and rare metals. None metallic minerals comprise dominantly gemstones. The exploration and mining activities in Vietnam have a long history of development. However, the mineral sector has not sustainably developed, due to the shortage of advanced knowledge, skill and expertise in research, exploration, extraction, processing and beneficiation of minerals as well as inefficient management.

Finally, Francis P. Gudyanga has presented a paper proposing to host a Pan-African minerals university of science and technology (PAMUST). PAMUST is proposed to be housed at the Scientific and Industrial Research and Development Centre (SIRDC) in Harare. SIRDC has enough land, modern research and development laboratories, a Library and documentation center, and other facilities.

Dr. Awni Al-Otoom

Dr. Mohammad Al-Harahsheh

Introduction

Availability of natural resources and resource-based industries is essential to prosperity and well-being of a country in this rapidly changing world. Many developing countries are generally richly endowed with minerals, biodiversity and other resources in comparison with developed countries. But these are often not fully exploited by them due to the lack of advanced skills, modern and innovative research and technology and the new world trade regime. African countries, for example, despite having large resources of platinum, diamond, gold, bauxite, etc., tend to lag behind their counterparts in the developing world and many minerals are just exported as ores, concentrates or metals, without significant downstream processing or value addition.

The minerals sector is vital for the creation of wealth, which immensely contributes to improving the overall quality of life, and is an important value-creating asset for developing countries. However, the common challenge among developing countries, including the member countries of the NAM S&T Centre, is to develop the mineral resources into products that can contribute to positive economic growth by circumventing the shortage of skill and expertise in specialised areas covering the entire minerals value chain - from extraction to products that can be used as components - or by taking appropriate measures like partnering among themselves. For this reason, there is a need for policy makers, scientists, technologists, academics and other industry experts to come together and assess ways and means to address the aforementioned skills gap.

In view of the above, the Centre for Science and Technology of the Non-Aligned and Other Developing Countries (NAM S&T Centre) organised a three days International Workshop on 'Minerals Processing and Beneficiation' during 25-27 September 2012 in Johannesburg, South Africa jointly with the Department of Science and Technology (DST), Republic of South Africa for brainstorming by international policy makers, scientists and technologists, mining industry executives, mineral industry and technology producers, investors, exploration people, mine owners and traders, planners and decision-makers. This workshop was attended by 63 participants from 22 countries, including Cambodia, Egypt, Eritrea, Guyana, India, Indonesia, Iraq, Jordan, Kenya, Malawi, Malaysia, Mauritius, Myanmar, Nigeria, Pakistan, Sri Lanka, Tanzania, Uganda, Vietnam, Zambia and Zimbabwe and the host country South Africa.

The present publication contains 17 scientific and technical papers presented at the above International Workshop and also some papers contributed by other invited experts on the subject. The contents of the book are divided into two parts: Section 1: 'Trade and Technology'; and Section 2: 'Policies and Strategies'.

I gratefully acknowledge the dynamic involvement and untiring efforts of Dr. Awni Al-Otoom, Director of the Energy Center and Dr. Mohammad Al-Harahsheh, Associate Professor, Department of Chemical Engineering of the Jordan University of Science and Technology, Irbid, Jordan for technical editing of this publication. Last, but not the least, I appreciate the valuable services rendered by the entire team of the NAM S&T Centre, in particular, Mr. M. Bandyopadhyay for overall supervision, and Ms. Isha Sehgal, Ms. Subhashree Basu and Mr. Pankaj Buttan in compiling the presented papers, liaising with the authors and editors, and giving a final shape to the volume.

I am sure that this book would serve as a useful reference material for the professionals working in the arena of mining and minerals and will help the developing countries in working out appropriate plans and strategies for the mineral resource exploitation and utilisation to generate wealth for them and prosperity for the masses.

Prof. Dr. Arun P. Kulshreshtha

Director and Executive Head,

NAM S&T Centre

Contents

Part II: Policies and Strategies

Part I

Trade and Technology

Chapter 1

The Status of The Gambia's Mining Industry: Processing and Beneficiation

Alieu Jawo

Geological Department, The Gambia,
Post Office Road, Kanifing, P.M.B 227, The Gambia
E-mail: alieujawo@hotmail.com

ABSTRACT

The Gambia is not known to be a mineral endowed country. In the 1950s, a British company, British Titanium, was briefly engaged in the exploitation of ilmenite.

Interest in The Gambia's mineral resources took a turn for the better in late 1990 to the beginning of 2000 when an Australian company revisited The Gambia's mineral potentials with particular interest on its heavy mineral sands.

The Gambia African Mining International Company, (GAMICO), a local company, is engaged in the mining of heavy beach sands. Mindful of Government's commitment of ensuring that mineral exploitation benefits the population, the company processes the sands in a two-step process before export. This consists of a primary and a secondary separation stages. In this way, opportunity is created for the employment of Gambians along the value chain. Government emphasizes linkages - forward, backward and lateral. The expected benefits are contribution to employment, skills development and thus poverty reduction. It may also be added that the companies pay rent and royalty to government which goes to government coffers. The Government of The Gambia, cognizant of this limitation, has endeavored to optimally utilize the available mineral resources through the efficient and optimal management of these limited mineral resources.

The Government of the Gambia is now working towards developing Regulatory framework designed to administer and optimize resource utilization. In order to alleviate the skills shortages, government encourages collaboration with local and international organizations and Institutions in addressing these shortages. To effectively develop the mineral resources into products that can contribute to economic growth and development.

***Keywords**: The Gambia, Titanium, Ilmenite, Zircon, Skills shortage, Processing, Separation plant.*

Introduction

The Gambia is located on the West- Coast of Africa, and bordering with Senegal to the east, north and south and the Atlantic Ocean to the west. It is one of the smallest main land countries with a total land area of the country approximately 11,000 square kilometers, and a population of about 1.8 million. The Gambia geologically is part of the Mauritania-Senegal-Gambia-Guinea Bissau-Guinea Conakry (MSGBC) coastal sedimentary basin.

The Gambia is not known to be endowed with high value minerals and mineral activities accounts for less than 5 per cent of the country's gross domestic product. The Government of The Gambia mindful of this fact has endeavored to optimally utilize the available mineral resources through effective and optimal exploitation. In line with this policy, minerals to be exported need to undergo further local processing and beneficiation before export. This would increase export value, and create employment opportunities for local.

The principal mineral resources for exportation from the 1950s to date are the heavy mineral sands. British Titanium Products Ltd. and Carnegie Corporation Ltd. had in the past operated in the Gambia and currently GAMICO Ltd. is the main company engaged in the exploitation of these resources. The company now does it primary and secondary processing in the country.

The quarrying for construction earth material is another sector that has over the years contributed to the economy of the country. The geology of the country is largely under explored. The quarrying for sand, laterite/gravel and clay take center stage when it comes to mineral resource exploitation in the Gambia. Traces of gold and tin have been discovered in the upper reaches in the river Gambia hence more exploration is needed in the country. The Government perceives the development of science and technology as the solution.

The availability of natural resources and resource-based industries is essential to prosperity and well-being of a country in this rapidly changing world. One way of achieving this target is moving away from being producers only, but by also be taking part in the processing and distribution of the mineral products.

Construction Earth Materials

In the Gambia the available resource database shows that the development path for the Mining Sector in The Gambia will be unique in Africa, and that the conventional approach adopted elsewhere in Africa will not be appropriate. Although the production and export of ore minerals is the target in many African countries, in The Gambia the emphasis is on the production and processing of construction earth materials to supply the internal construction market. To achieve this development the Government has implemented measures aimed at stimulating the demand side for domestic construction materials, the establishment of an internal investment regime through appropriate legislation, and the stimulation of the supply side through the provision of easy access to the materials. Mineral exploration and quarrying are included in the list of priority sectors eligible for special investment incentives.

The mineral activities in this sector consist essentially of a number of very basic, informal, and artisanal ventures. It involves individuals or small companies getting few trucks and machinery for the extraction of earth material for use in construction and building. The main focus of these activities is on sand quarrying, gravel quarrying, cockleshell and limited clay extraction.

A high percentage of Gambians involved in mineral resource exploitation activities are in this sector. This sector plays a significant role in the mineral industry.

Heavy Mineral Sands

In The Gambia the three main proven resource areas for mining of ilmenite, zircon and rutile were identified in the coastal region of The Gambia (Figure 1.1) at Sanyang, Batukunku and Kartung (Whyte and Russel, 1988). However, these deposits are relatively small on an international scale.

The Gambia Government, supported by the Commonwealth and the UN, initiated several exploration and mining feasibility studies in the 1970- 1990s. The Geological Department with the support of these International bodies and other international consultants including Mathew Hall Ortech Ltd., UK, evaluated the mineral sands deposits. However the deposits were regarded as sub-economic at the time (Ortech, 1976). In the late 1990s- 2003 this situation changed, when international companies rekindled their interest in the Gambia mineral sands. The Gambia Government in 1990s made considerable efforts to attract investments for developing the economy.

The heavy mineral sands deposits are located within Quaternary coastal zone in a series of well identifiable raised ancient beach dunes generally parallel to the current coast line. The heavy mineral deposits consist of high to medium grade and easily accessible. Total reserves were reported at 1 million tons of heavy minerals with a potential to increase to a maximum of 0.50 million tons of Heavy mineral in (Barcaly, 1997). The mineralogical suite contains a high ratio of valuable heavy minerals of which 71 per cent, ilmenite 3.3 per cent rutile and 15 per cent zircon.

Mining of Heavy Mineral Sands

In The Gambia mineral exploration in the coastal areas was started in 1950s by British Titanium Products Ltd. (BTP) to substitute their disrupt supply of titanium feedstock from India (Ortech 1976a). In 1965, mining operations began at the Batukunku deposit, following the discovered of three placer deposits in Sanyang, Batukunku, and Kartung. It is reported that about 60,000 tons of ilmenite and rutile were exported by (BTP) to the UK during the period from 1956 to1959 before the closure of the operations due to the resumption of the Indian supply (Whyte and Russel. 1988). As a result this sector was inactive from 1959 to late 1990s, however,

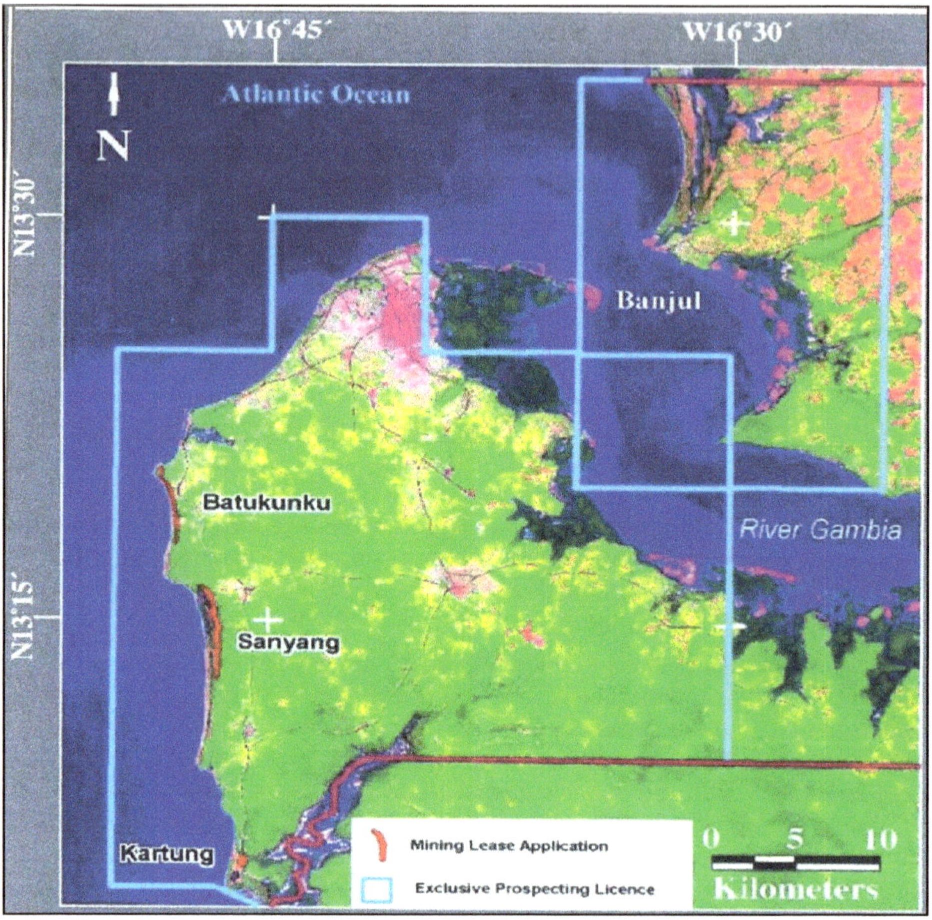

Figure 1.1: Map Showing the Heavy Mineral Sand Deposits of The Gambia (Source from Carnegie Minerals Gambia Limited, 2005).

for Gambians, the boom in the mining of construction earth materials such as sand, gravel, and clay serves as substitute.

Interest in The Gambia's mineral resources took a turn for the better in the late 1990 to the beginning of 2000 when an Australian company revisited The Gambia's mineral potentials with particular interest on its heavy mineral sands. The company, Carnegie Corporation Ltd reviewed all available geological and economic data. It was demonstrated that both titanium minerals and zircon products from the Gambian mineral sands resources could be potentially saleable at a number of markets. The company in the early 2000s secured a mining licence and undertakes only partial processing of the material in the country. The concentrates was then ship for further processing at facilities already established in China.

In 2007 GAMICO Co Ltd entered the Gambia's mineral sands industry and continues the partial processing of the materials and shipping the concentrates for further processing at facilities elsewhere in the World. However this trend has since change and now the company does its primary and secondary processing in the Gambia before exporting the finished products. This has increased the level of employment, and adds value to the goods, hence increase government revenue.

Processing and Beneficiation of the Heavy Mineral Sands

The lack of technical skills and technology in The Gambia, coupled with the perception that mineral sands deposits were too small to support onsite processing and beneficiation, most mining companies set their eye on processing these minerals abroad. To overcome these perceived impediments, mining companies propose to undertake only partial processing of the material in the country and to ship the resulting concentrate for further processing at facilities already established elsewhere, notably China.

Carnegie Cooperate Ltd., for instant during its feasibility studies identified Astron Limited for partnership, an Australian company with mineral processing plant in China. This partnership was reached in order to ensure a continuous supply of raw materials for Astron Ltd. Carnegie Cooperate Ltd initial economic evaluation of the total size of the Gambia mineral sands resources does not support the commercial development of an onsite TiO_2 processing plant.

Primary Processing

In 2002-2003 when Carnegie Corporation Ltd. started mineral exploitation the company proposed to conduct trial and test the market potential of The Gambian ilmenite. Subject to the success of the trail, the company commenced primary processing operations to produce preliminary mineral sands concentrates in The Gambia.

In its mineral processing, Carnegie Corporation Ltd. adopts both wet (dredge mining) and conventional dry mining techniques feeding a conventional central processing unit employing separation techniques. Figure 1.2 shows the simplified flow diagram for the dredge and floating concentrator plant. The plant produces about 1.8 tones per hour of the crude heavy mineral concentrate that contains ilmenite, zircon and rutile (no less than 40 per cent TiO_2 and 10 per cent ZrO_2 in total). The low grade of the crude concentrate usually makes it uneconomic to transport the concentrate over long distance.

Secondary Processing and Beneficiation of Mineral Sands

In line with Government's demand and desire that further processing of the mineral sands be done in the Gambia, to create jobs, enable skills transfer and generate more revenue for government, GAMICO Co Ltd., has been working on this plan. The Company in 2010 contracted a South African company to build a magnetic separation plant at its Sanyang site and the plant was commissioned at the beginning of the year. The concentrate instead of being send to China for further processing is being process in the Gambia. The Separation Plant separate the concentrate into magnetic (ilmenite) and non-magnetic (mostly zircon).

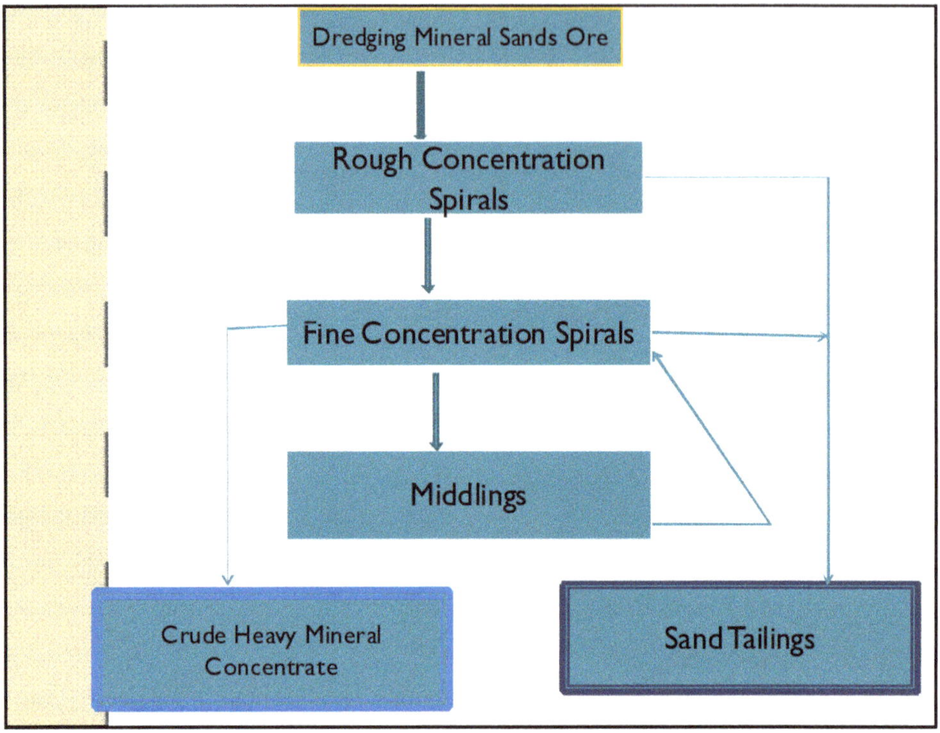

Figure 1.2: Simplified Process Flowsheet for Dredge and Floating Concentrator Plant in Sanyang.

Discussion

In resource-rich but economically-poor countries, the mining sector can enhance socio-economic development and reduce poverty. The Gambia despite not being endowed with mineral resources is formulating policies geared towards maximizing the benefits from mining. The Processing of minerals locally creates employment opportunities for local communities (Figure 1.3) and increase export value of the mineral resources.

The Gambia, shares the Africa Mining Vision, as well as the ECOWAS Directive on Mining. The Mining Regulations (under formulation) of the country places emphasis on companies employing Gambians in certain technical areas, however due to skills shortages most of such positions are filled by expatriates. Most of The Gambians involved in the mining sector are un-skills workers see (figure3) engaged in labour intensive and manual jobs.

The expected benefits of further processing as stipulated above includes contribution to employment, skills development and thus poverty reduction.With the establishment of a Separation Plant by GAMICO Co. Ltd., now Gambians were trained to operate the Separation Plant. This creates employment opportunities for Gambians.

Figure 1.3: Gambians Laborers Loading and Transporting Concentrate in Bulker Bags at Brufut.

The growth potential of the Gambian mineral sector will depend on further mineral exploration and currently an international company, Pan African Gambia Ltd is licensed to prospect for minerals in The Gambia.

Conclusion

The mineral resource sector of The Gambia includes exploring, extracting and processing resources such as ilmenite, zircon, rutile (heavy mineral sands), sand, gravel and clay. The Gambia shares the mining vision of African countries of equitable and optimal exploitation of mineral resources to underpin broad-based socio-economic development and sustainability.

The Gambia aimed to foster fair dealing with mining companies and encourages training and skills transfer. Mining companies in The Gambia implement on-the-job training programmes and give preference to the employment of nationals of the Gambia who have appropriate qualifications, and this has boosted local employment. The Government is now promoting building capacity by encouraging companies to train Gambians on mining related training; this is aimed at addressing the skill shortage in the Gambia. The Gambia aim to promote its construction earth material industry as it plays a significant role in the country's socio-economic development. Quarrying has been a driving force for much of the exploration and it's a lucrative business for quarrying operators.

The commissioning of a Separation Plant by GAMICO Co. Ltd. marks an important milestone in the history of mineral processing in The Gambia. These developments create employment opportunities for Gambians and add value to the country's mineral resources.

Acknowledgement

First of all, I wish to thank the Geological Department of The Gambia and The Gambia's Ministry of Higher Education, Research, Science and Technology for the support given in the preparation of this country report. I am highly indebted to the Department of Science and Technology of South Africa and the Centre for Science and Technology of the Non-Aligned and other Developing Countries for organizing the international Workshop on Mineral Processing and Beneficiation.

References

1. Barclay, J. (1997). Gambian Titaniferous Beach Sands. Report to the UN Industrial Development Organization.
2. Matthew Hall Ortech Ltd. (1976). Investigation on a Zircon Dump and Virgin Titaniferous Beach Sands in The Gambia. UK.
3. Whyte, W. J., Russel, T. S. (1988). Geology and Mineral Reserves of The Gambia. Geological Unit of the Ministry for Local Government and Lands of The Gambia.
4. TZ Minerals International Pty Ltd. (2000). Flowsheets for the Brufut Zircon Stockpile and The Gambian heavy mineral resource.

Chapter 2

Instrumental Analysis and Classification of Ekon Kaolin

K. Akufo-Kumi, H.A. Obiri, G.N. Laryea and W.O. Oduro

Institute of Industrial Research, CSIR,
P.O. Box LG 576, Legon, Accra
E-mail: kakufokumi@gmail.com

ABSTRACT

Nuclear-related analytical techniques (such as energy-dispersive X-ray fluorescence, neutron activation) and atomic absorption spectroscopy analyses were employed to find the elemental concentrations of Si, Al, Fe, Ti, Na, Ca, Mg and K and their oxides in Ekon kaolin samples. The mineralogy of the samples was determined by X-ray diffraction analysis. Silica and alumina contents of samples ranged between 2 per cent -95 per cent and 0.85 per cent - 58.9 per cent, respectively. The assay of the raw material shows a percentage purity of about 84 per cent with the impurity constituting allowable limits of ferrous materials and other flux materials concentrations. The Ekon Kaolin had alumina and silica contents of about 38.82 per cent and 44.72 per cent, respectively and was found to have properties that make it a potential raw material for the production of alumino-silicate refractories suitable for non-ferrous industries.

***Keywords**: Kaolinite, Chemical and mineralogical analysis, Raw material assay.*

Introduction

Refractories are ceramic products manufactured to withstand high temperature corrosive conditions encountered in industrial furnaces *i.e.* temperatures of above 1000°C corresponding to cone number 18 (Gil *et al.*, 1989). Ceramic products must have high melting temperatures, good high temperature strength, chemical inertness, resistance to abrasion, and specific or tailored thermal properties (CAEA Vienna, 1992). These materials are used to produce furnace liners, crucibles, molten metal ladles and a host of other special temperature products.

There are several refractory product groups, each named after the main oxide component. Some of these are silica, fireclay, high alumina, chrome-magnesite, refractory bricks etc. Refractories are manufactured from a variety of ceramic raw materials including clay with high alumina content, bauxite and relatively pure oxides such as alumina, silica, magnesite, chrome ore, etc. The main divisions of refractories are fireclay, silica, alumina and basic pure oxide refractories, etc. (Anderson, 1994). The properties of refractory products depend on the principal minerals (the major crystalline phase having a high melting point and the presence of small amounts of mineral phase of lower melting points which reduce the refractoriness of the product.) and their distribution. The chemical stability depends on the mineral assemblage presented in the product.

High purity raw materials are preferred in order to decrease the amount of low melting minor constituents and to avoid the formation of low viscosity eutectic in the brick (AMTM, Vienna, 1991). Numerous raw materials containing Al_2O_3 and SiO_2 as major components have refractory properties. The fireclay products represent the greatest part of this group of refractories. The classifications of these products are according to their Al_2O_3 content. Clay consists of different amounts of various clay minerals and fine-grained quartz. When fired at 1200-1400 °C, it leads to the formation of mullite and SiO_2 which subsequently then become dehydrated. Mullite with a melting point above 1800 °C does not exist in nature in minable quantities. It is therefore produced synthetically. Their crystals are characteristically needle-like in shape. Generally, there are two types of mullite materials on the market; the synthetic and the electro-fused quality (Katsaki, 1989). Bauxite, alumina, kaolin and quartz sand are the raw materials mostly used to synthesize mullite. The electro-fused quality made of alumina and high purity kaolin contains more than 95 per cent of mullite, the rest being a mixture of corundum and glass. The sintered quality contains between 85-90 per cent mullite with alumina content of 71-73 per cent. In the synthesis of mullite, its composition depends on the properties of alumina, silica and on temperature schedule (Hihara &Kawamoto, 1991). The mullite, $3Al_2O_3.2SiO_2$, was found to be the only stable phase whereas $2Al_2O_3.SiO_2$ was described as a mullite-corundum composition. No evidence is known of solid solution between both alumino-silicates.

Indicated in Scheme 1 is the chemical reaction of kaolinite and the phase changes it undergoes during heating.

These phases are the main constituents of the fired clays but because the refractoriness and chemical resistance of fireclay products are frequently not adequate to stand the increasing severe conditions in many industrial furnaces, other refractory products with high alumina content as well as mullite are preferred (Kamizawa &Yamamoto, 1991).

To assess the refractoriness of local raw materials and their industrial usefulness, the percentages of the oxides like Al_2O_3, SiO_2, Fe_2O_3, TiO_2, K_2O Na_2O, and MgO must

$Al_2O_3 \cdot 2SiO_2 \cdot 2H_2O$ (Kaolinite, Kaolinite -**Meta Kaolinite**)

↓ Dehydration at 500 °C- 600 °C

$6(Al_2O_3 \cdot 2SiO_2)$ **(Meta Kaolin)**

↓ Exothermic Peak at 980 °C

$3(Si_3Al_4O_{12})$ (Al-Si Spinel)

↓ 1100 °C

$2(3Al_2O_3 \cdot SiO_2) + 5SiO_2$ (Mullite - Cristobalite)

Scheme 1: Schematic Representation of the Chemical Reaction of the Kaolinite and the Phase Changes during Heating.

be known to enable the grading of the product (Bashir, 1994). Consumer refractories in Ghana are in the group of alumino-silicates. The demand for these has increased rapidly lately due to Government policy on local content in industrial production programmes. This is intended to ease pressure on the country's scarcely available foreign exchange. There is therefore, the need for the study and classification of Ghana's bauxite reserves which would lead to the development and production of local substitute for alumina-silicate refractories. The evaluation of raw materials is a principal requirement in any economical production process. The comparison of the quality potential of the raw material with the market demand (products specifications) and the determination of the effort that are required under technical, social, environmental and economical aspects are pre-requisites in industrial development (Osae, 1988). It is not sufficient to have a raw material source and to know about the extent to this deposit. It is essential to study the properties, the variations of coverage quality and to study the market potential of it before the source is exploited. The evaluation of the raw material is a true passing requirement from the early state of recognizing and planning of the depletion of the natural resources. Changes in both on site and market qualities lead to adjustments of the manufacturing process by either decreasing the product line or causing deviation into other market areas.

In this study, multi-elemental and mineralogical analyses were conducted on locally available Ekon Kaolin which is in the reserve of 15-20 million metric tons in Ghana (Kesse, 1985) to determine its suitability for refractories applications.

Materials and Experimental Method

Sample Collection

Clay samples of massive appearance were collected randomly over a 0.5 km^2 geographical area from Ekon in the Central Region of Ghana using a Macintosh pitting set-up to sample at different depths. The samples were mixed thoroughly and a 2 kg split sample was taken for laboratory analysis. The laboratory sample was dried in an open air oven at 110 °C for two hours and stored in a desiccator.

Sample Preparation and Irradiation

About 500 grams of the sample was washed (to remove coarse sand and organic matter) and dried. It was pulverized and ball milled for particles to pass through a sieve size 0.6 mm. Infinitely-thick and intermediate samples of 4.9 cm^2 cross-sectional area were prepared using a hydraulic press (SPECAC) with a pressure of 10 tons.

Standard pure oxides of Ca, Zn, As, Ti, Mg, Al, Cu, Cr, V and Ni were also taken and ground using an agate mortar and pestle to equivalent particle sizes as the Ekon Kaolin sample.

Samples were irradiated for 20 minutes to obtain the spectra. Ten standard samples were also irradiated for 10 minutes each. The results obtained from the standard samples were used for both qualitative and quantitative analysis of the samples using the x-ray spectrometer (Hihara and Kawamoto, 1991). It is worthy to note that for quantitative purposes, the calibration was normalized to counts per second. The results were therefore, not affected by difference in irradiation times.

Instrumental Neutron Activation Analysis (INAA)

Elemental analysis of the ceramic raw material samples was also done using neutron activation analysis. The nuclear-related analytical technique was able to detect the element with lower atomic numbers, which otherwise, the x-ray fluorescence analysis could not detect. Ceramic raw materials were pulverized and about 150 mg of sample was weighed and wrapped in polyethylene. This was then sealed in polyethylene capsules and sent to the reactor by means of a rabbit system. Parameters such as neutron flux, irradiation time, decay time and counting time were carefully selected depending on the half-life of the element of interest. A gamma monitor was used to ascertain its readiness for counting on the gamma detector. Energy calibration was done by collecting spectra of standard elements. The objective was to convert count per channel to count per energy, and thus obtain a plot of energy (KeV) against channel number (Longerich, 1995).

Determination of Silicon Content in Sample Using Atomic Absorption Spectrophotometer (AAS)

The sample wet digested for analysis with AAS by high temperature fusion by weighing 2.0 g of sample with 5.0 g of a 1:1 ratio mixture of sodium carbonate/ sodium tetra borate flux before heating to 600 °C followed by dissolution with 50 per cent HCl. The flame technique was used during the AAS analysis, employing a nitrous oxide/acetylene flame. The percentage Si composition of the sample was determined after calibrating the AAS with known concentrations of standards.

Results and Discussion

Chemical Analysis

Energy Dispersive X-Ray fluorescence, instrumental neutron activation and atomic absorption spectrophotometer analyses of Ekon Kaolin samples revealed the presence Al, Si, Fe, Mg, Ti, Ca, Na, and K which is typical of kaolin samples.

The average elemental concentration in Ekon Kaolin determined using the different analytical techniques; EDXRFA, INAA and AAS are tabulated in Table 2.1.

Table 2.1: The Average Elemental Composition of Ekon Kaolin using different Analytical Techniques

Sample (per cent)	*Si*	*Al*	*Fe*	*Ti*	*Ca*	*Mg*	*Na*	*K*
Ekon Kaolin	21.48± 0.88	20.54± 0.33	0.39± 0.03	0.82± 0.03	0.15± 0.00	0.22± 0.01	0.01± 0.00	0.64± 0.05

The chemical compositions of the oxides of the elements determined from the nuclear analytical techniques were derived from the Equation 1 and the results are as shown in Table 2.2.

$$\%\,Oxide = \frac{Molecular\ wt.of\ Oxide \times Conc.of\ Element}{Atomic\ wt.of\ Element \times n} \tag{1}$$

where, n is the number of occurrence of the element of interest in its oxide

Thus the percentage composition of Al_2O_3 in the Ekon Kaolin as represented in the elemental analysis as Al is

$$\%\,Al_2O_3 = \frac{101.96 \times 20.54}{26.982 \times 2} = 38.82$$

The percentage loss by weight often called loss on ignition (LOI) when sample was gradually heated to a temperature of 1000 °C was obtained using thermo gravimetric analysis. The sum total of the oxides and the loss on ignition valve are tabulated in Table 2.2.

Table 2.2: The Percentage Oxide Composition of Ekon Kaolin Derived from Elemental Concentrations and Molecular Weights

Sample (per cent)	SiO_2	Al_2O_3	Fe_2O_3	TiO_2	Na_2O	CaO	K_2O	MgO	LO I	Total (per cent)
Ekon Kaolin	44.72	38.82	0.57	1.37	0.01	0.21	0.77	0.36	13.19	100.02

The qualitative and quantitative analysis of Ekon kaolin using complementary analytical techniques and the computation of their oxide composition indicated the presence of alumina to silica ratio of about 39 per cent : 45 per cent. The proportions of alumina and silica make Ekon Kaolin of high ceramic interest especially in the development and manufacture of mullite and high alumina alumino-silicate refractory bricks.

Mineralogical Studies on Ekon Kaolin by XRD and DTA

The mineralogical study of the sample of Ekon Kaolin by XRD indicated the mineral Kaolin as dominant in the material as shown in the different phases by intense peaks at 2θ values of 14° and 37° in Figure 2.1. The X-ray diffraction test

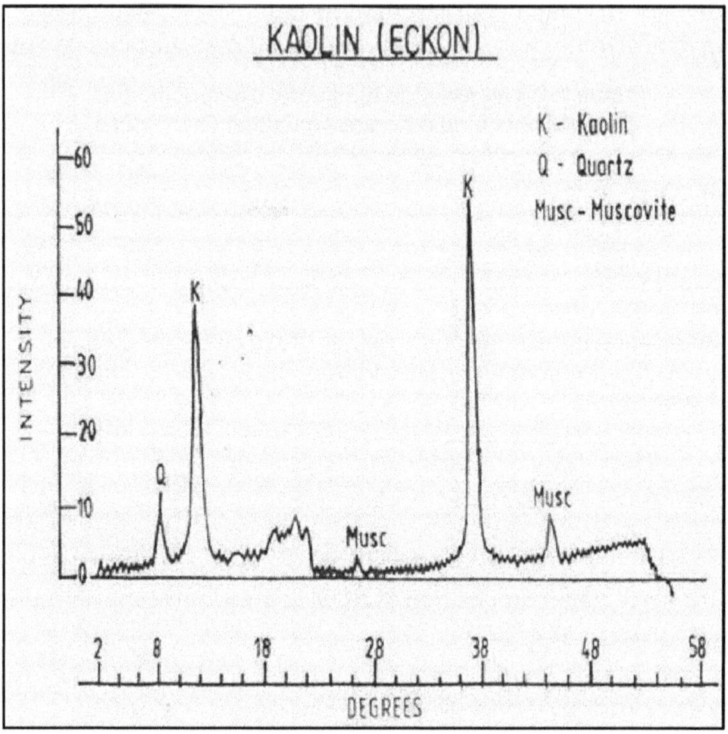

Figure 2.1: An X-ray Diffractogram of Ekon Kaolin Showing Kaolin (K) as the Dominant Mineral and Some Quantities of Quartz (Q) and Muscovite (Musc).

showed high peaks of kaolin ($6Al_2O_3.2SiO_2$)) and some peaks of muscovite and quartz which were quite minimal.

The differential thermal analysis of the sample showed the thermal transformation patterns of kaolinite as indicative by endothermic peaks around 600°C (Gunicheva *et al.*, 1995) in Ekon Kaolin as.shown in differential thermal curve in Figure 2.2.

Conclusions

The energy-dispersive X-ray fluorescence, instrumental neutron activation and atomic absorption spectrophotometer analyses indicated the presence of the elements of interest on the sample in different concentrations. There were other elements which were in traces in the order of concentration of parts per billion, which were not of interest and not in measurable quantities. From the average concentrations of elements obtained by the analytical techniques employed, the oxides of the elements were determined based on their molecular and atomic weights and the ratio of alumina to silica was determined to be 39 : 45. Hence Ekon kaolin has alumina/silica ratios which are within the range for the development of fireclay refractories if proper treated.

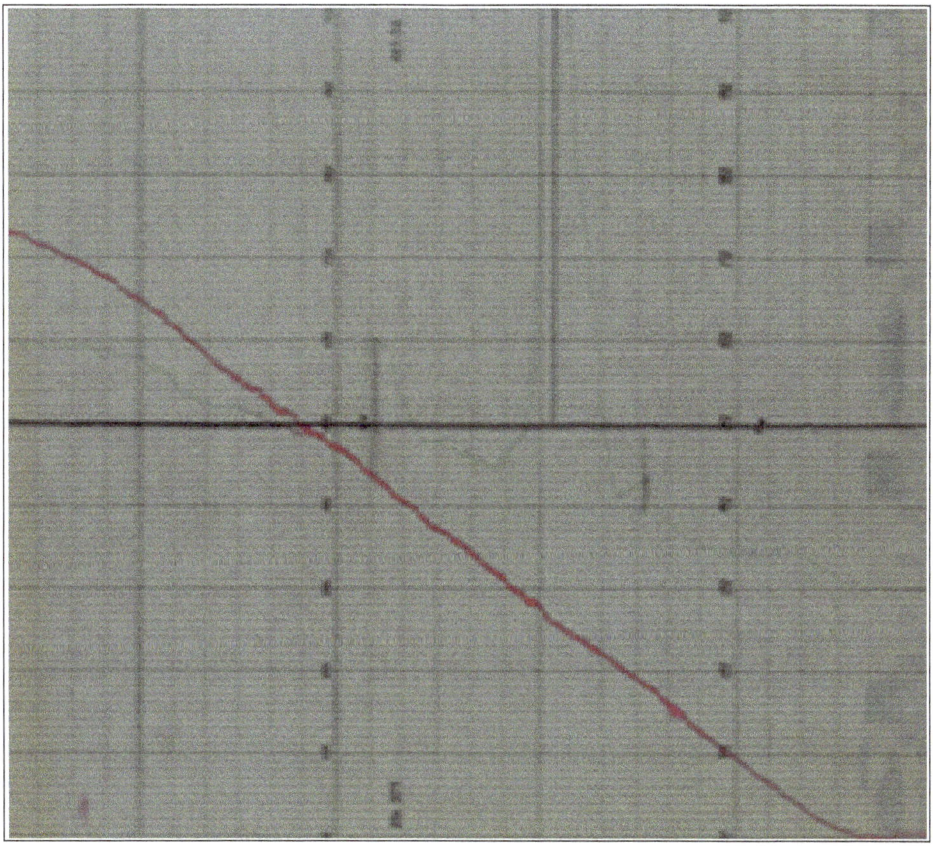

Figure 2.2: The Differential Thermal Curve of Ekon Kaolin Depicting Intense Peaks for Kaolinite as the Major Component.

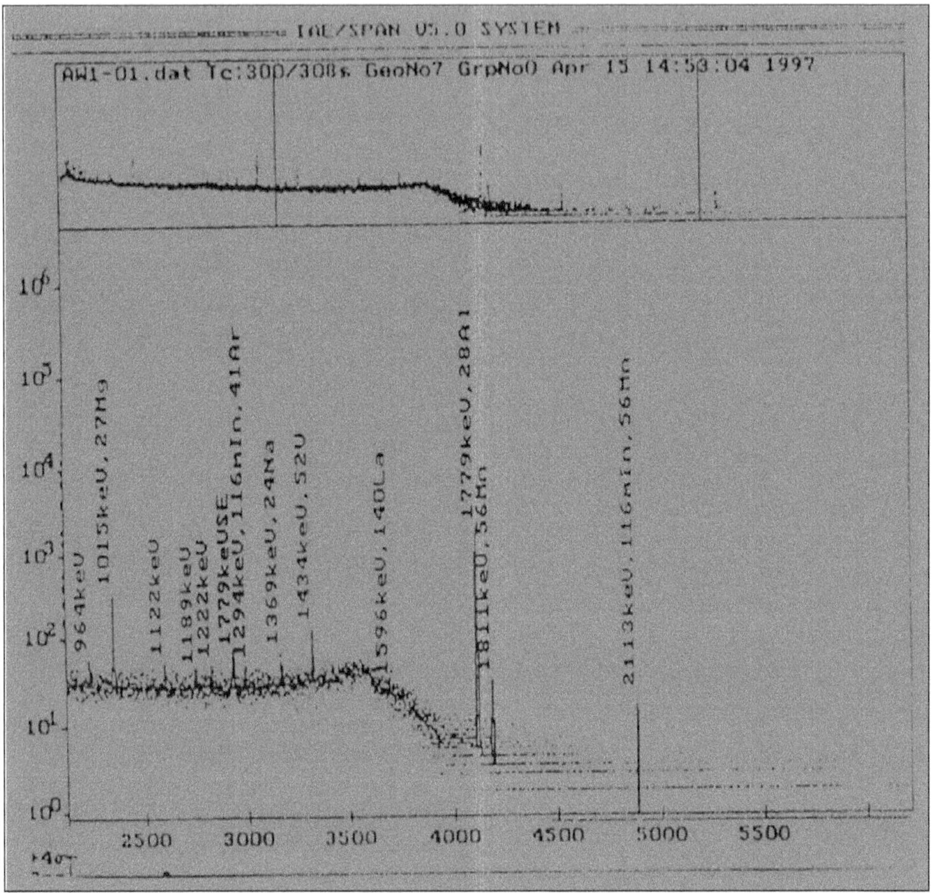

Figure 2.3: A Neutron Activation Analysis Spectrum of the Sample of Ekon Kaolin as Complementary Elemental Analysis Technique to the X-Ray Fluorescence.

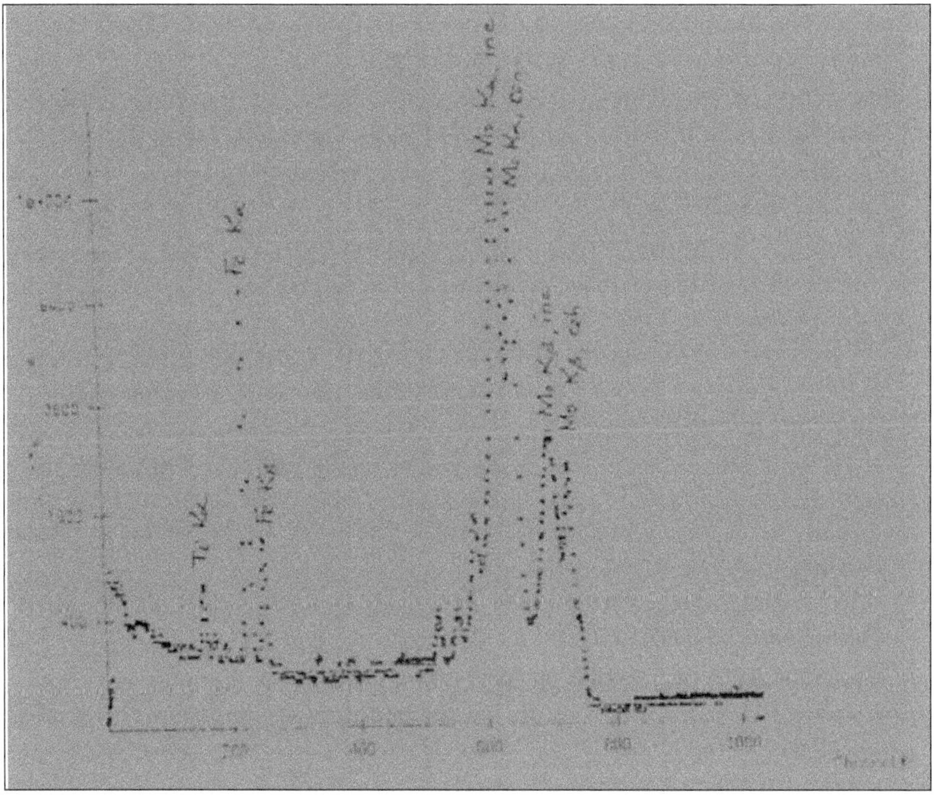

Figure 2.4: An XRF Spectrum of Ekon Kaolin for Complementary Elemental Analysis. The quantitative and qualitative analysis were conducted with reference to standards of known purity.

Acknowledgements

The authors are grateful to the National Nuclear Research Institute and the Chemistry Department (Ghana Atomic Energy Commission) and CSIR Institute of Industrial Research.

References

1. ANDERSON, L. (1994). Development of the Refractories Production from Local Raw Materials and in Sub-Regional Cooperation, Evaluation of Local Raw Materials for Refractories Production, Technical Report.
2. BASHIR, I. B. (1994). Refractories Production and Practice of Nigeria Federal Institute of Industrial Research, Oshodi, Lagos, Nigeria, Technical Report.
3. C.A.E.A. Vienna (1992). A Handbook on Nuclear Activation Data, Technical Reports Series No. 340.
4. GIL, F. B., BARREIRA G., GUERRA M. F. and ALVES L. C. (1989). Quantitative Elemental Analysis of Thick Samples by X.R.F. and P I X E, X-Ray spectrometry pp. 157-164.
5. GUNICHEVA, T. N., AISVEVA, T. S. and ATOMIC, V. P. (1995). Non-Destructive X-ray Fluorescence Analysis of soils and Marine sediments, X-ray Spectrometry, 24(4), pp. 187-192.
6. High Temperature Ceramics, Advances in Materials Technology Monitor (1991). Issue No. 23, compiled by the Industrial Technology Development Division, Department for Industrial Promotion, Consultations and Technology, Vienna, Austria.
7. HIHARA, M. and KAWAMOTO J. (1991). Composite Mullite, Osaka University. Techno Japan 24, 80.
8. KAMIZAWA, J. and YAMAMOTO T. (1991). Ultra Fine Particles of Alumina, Faculty of Chemical Engineering, Tokyo Institute of Technology; Techno Japan 24, 68, pp. 37-39.
9. KATSAKI, T. (1989). Mullite of Near Theoretical Composition, Techno Japan 22 (8), 78.
10. KESSE, G. O. (1985).The Mineral and Rock Resources of Ghana, A.A. Balkema/ Rotterdam/Boston, pp. 416-437, 581.
11. LONGERICH, H. P. (1995). Analysis of Pressed Pellets of Geological Samples Using Wave length. Dispersive X-ray Fluorescence Spectrometry; X-ray Spectrometry 24(3), pp. 123-136.
12. MACKENZIE, R. C. (1970). Differential Thermal Analysis, Academic Press, London and New York, pp. 498-499, 524-525.
13. OSAE, E. K. (1988). The Principles of Neutron Activation Analysis (Unpublished).

Chapter 3

Initiatives by the Guyana Geology and Mines Commission to Improve Mineral Recovery, and Increase Profitability in the Local Mining Sector: A Status Update

Dianne McDonald

Head-Mineral Processing Unit,
Guyana Geology and Mines Commission, Brickdam, Georgetown, Guyana
E-mail: mmmdggmc@yahoo.com

ABSTRACT

The Mining Industry in Guyana is made up of large, medium and small scale sectors, which are regulated by the Guyana Geology and Mines Commission (GGMC) according to the Mining Act of 1989. The medium and small scale sectors are the largest, by virtue of the number of operations, employment, earnings and contribution to the local economy. The traditional minerals produced are bauxite, gold, and diamonds, with bauxite being exploited at the large scale, while gold and diamonds are exploited at medium and small scales. Improvement of mineral commodity prices on the world market has resulted over the last three years in greater foreign exchange earnings from gold and diamond mining operations. In 2011, this sector contributed to over 75 per cent of the input of the Mining Industry to the GDP of Guyana (the industry's overall contribution was 11 per cent). Higher mineral prices have attracted more entrants to small and medium scale sectors, and while the positive contribution to Guyana's economy cannot be denied, the environmental footprint of this sector is a factor that has to be strategically addressed.

Generally, it is agreed that small and medium scale operators in Guyana have improved their productive capacity over the years. They have moved from the simple pick axe and shovel to gravel pumps and excavators which allow them to move tonnes of material per day. However, the mineral recovery methods generally remain the sluice box and batel for gold, and the lavador (Brazilian-jig) and hand-sieves for diamonds, where recovery is often less than 50 per cent. Improvement in mineral recovery has to be addressed because of the environmental management requirements by law; the national requirement to exploit mineral resources in a Low Carbon Development environment; and the imminent international ban on mercury trade.

The GGMC as part of it mandate to support and promote the development of the Small and Medium Scale mining sector, has taken a number of initiatives. These are: provide assistance to the mining sector to improve mineral recovery for increased efficiency and to boost profitability; increase mineral products base beyond the traditional gold, diamonds, and bauxite; develop value added mineral products; and improve environmental and occupational safety and health management in the sector. Additional initiatives include improving current practises; promoting improved recovery methods by demonstrations; promoting mercury free methods for gold extraction; educational and awareness campaigns by seminars, brochures, flyers; technical assistance campaigns to mine operators. A mineral processing research and development facility has been developed to support these initiatives; also it will be a requirement by law for mine operations to have trained and certified managers running their operation.

Keywords: *Improve mineral, Recovery, Profitability, Promoting, Mercury-free, Initiatives, and Products.*

Introduction

The Mining Industry in Guyana is regulated by the Guyana Geology and Mines Commission (GGMC) which was established in 1979 as a successor to the Geological Surveys and Mines Department of the Ministry of Energy and Mines [1]. This semi-autonomous agency, established to administer, regulate, promote and provide technical direction for the development of Guyana's mineral resources and subsequently, petroleum resources, is governed by the GGMC Act of 1979 and administers the Mining Act of 1989. Within this framework, GGMC retains a portion of the fees and royalties it collects and over the years, with recent improving gold prices, has developed into a financially strong institution with the ability to support technical development towards the improvement of mineral recovery in the Guyana small and medium scale mining sector.

The Mineral Processing Unit (MPU) was set up in 1989 to support the GGMC's technical programmes for promoting best practices in mineral recovery among Small and Medium Scale mine operators. Under the GGMC's current five (5) years plan (2008-2012), the MPU is supporting GGMC's vision of 'Unlocking the mineral and petroleum wealth of Guyana', under the following strategic objectives:

- To promote and support increased and diversified mineral development
- Conduct mineral processing test work
- Test alternative and new mineral recovery systems

- ✰ Promote enhanced and environmentally safe mineral recovery techniques
- ✰ To recruit and train additional staff needed to complete the work programme

Over the years, through education and awareness activities, equipment selection, modification, testing, practical demonstrations of mineral processing equipment and techniques and training, the GGMC's initiatives have continued and are ongoing. Within the last two years, the GGMC made the bold move to set up mineral processing research and development laboratory and pilot facilities, with the aim of equipping the GGMC with the necessary capacity to pre-concentrate gold ores using various equipment, and introduce and effectively manage the safe (for both users and the environment) use of Cyanide Leach Technology for gold recovery in medium scale operations. In particular, under this initiative, GGMC officials were trained, practically and theoretically, as Trainers. The GGMC also purchased two complete gravity processing plants to demonstrate improved gravity recovery methods to small and medium scale miners.

These are all on-going activities, and there are challenges which include building capacity within GGMC to aid the transfer of skills to the small and medium scale sectors, which would support the acceptance, setting up, testing, fine-tuning, and adopting improved mineral recovery by operators. The overall goal is to promote efficiency in traditional practices and introduce improved methods that would improve profitability, increase minerals products, and promote new mineral based industries in the local economy. Another challenge is the need is for small and medium scale operators to operate within the framework of the national Low Carbon Development Strategy that has been adumbrated for Guyana, which requires a higher level of compliance with key environmental regulations and strict monitoring of deforestation at a national level.

The Mining Industry in Guyana consists of large, medium and small scale sectors. Under the Mining Act, state lands in the six Mining Districts can be allocated for Small, Medium, or Large Scale prospecting or mining;

or for Reconnaissance Scale geological, geochemical or geophysical surveys over large areas. The corresponding mineral properties are Small Scale Claims (max. size is 800ft x 1500ft or 27.5 acres) or River Locations under Claim or River Location Licences for mining; Medium Scale Prospecting or Mining blocks (150 and 1200 acres) under Prospecting or Mining Permits; Large Scale prospecting and Mining properties under Prospecting or Mining Licences and Reconnaissance Scale properties under Permissions granted for same. Operationally, small scale mining is defined as a mining occurring on a Claim or River Location from which a volume is excess of $20m^3$ but less than $200m^3$ of material, inclusive of any overburden, is excavated or processed as an aggregate in any continuous twenty-four hour period; medium scale mining occurs on a Mining Permit block from which a volume in excess of $200m^3$ but less than $1,000m^3$ of material is handled as described above within the same time period; and the large scale mining occurs on Mining Licence property from which a minimum volume in excess of $1000m^3$ of material is handled [2]. In this discourse, these are being referenced as it relates to mineral recovery. In Guyana, the medium

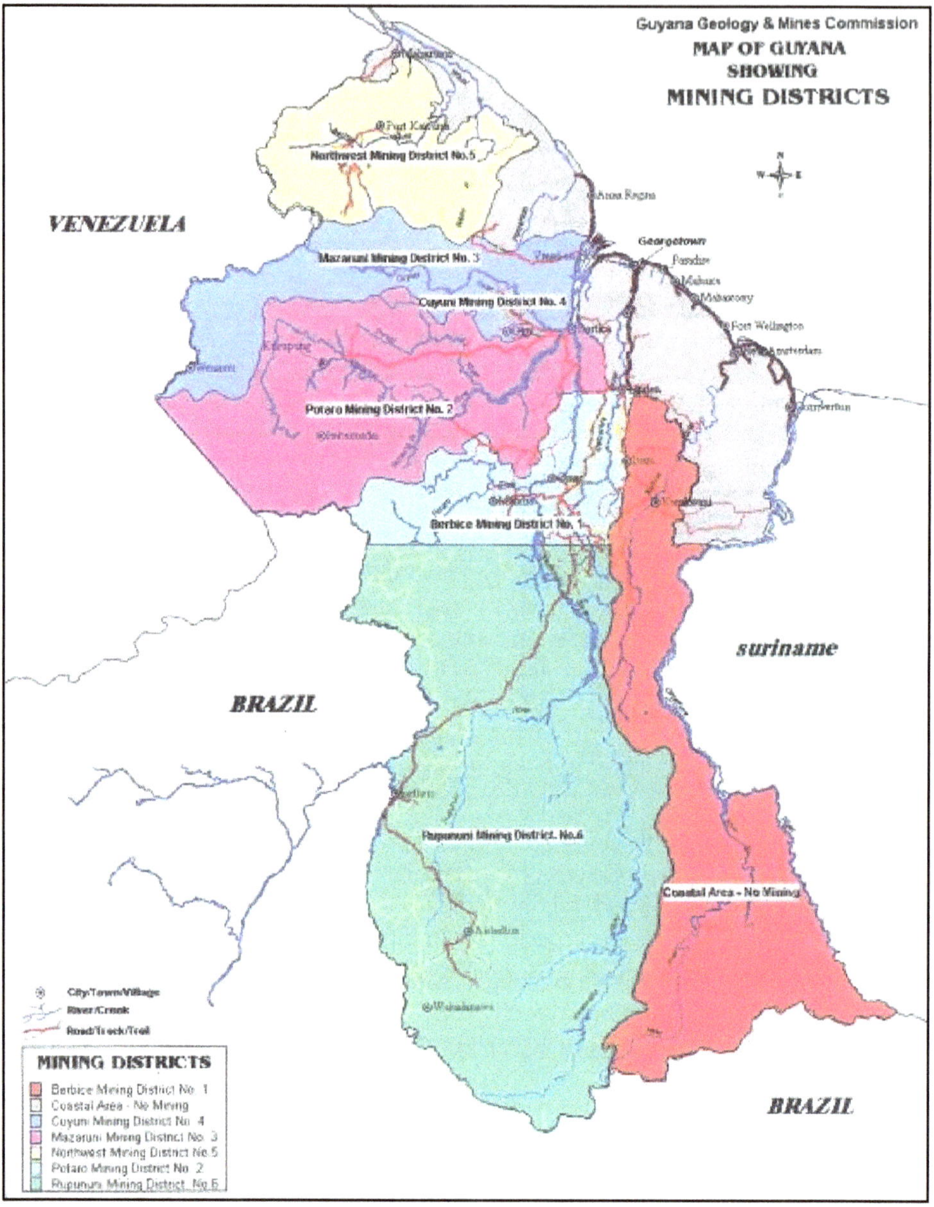

Figure 3.1: Map of Guyana's Mining Districts.

and small scale operations are the largest, by virtue, in the number of operations, employment, earnings and contribution to the local economy.

All mining operations are currently privately owned. Bauxite products are dried bauxite of refractory, chemical and metal grades. Alumina was produced from 1961 to 1981. Historically, the bauxite sector independently developed and upgraded

Table 3.1: Contributions of the Sectors to Guyana's Mineral Production Value (2006 –2011) (per cent).

Sector	*2006*	*2007*	*2008*	*2009*	*2010*	*2011*
Small and Medium Scale sector (gold and diamonds)	69	64.3	66.2	77.8	76.6	80.3
Large scale sector (bauxite)	28.6	31.8	29.7	18.5	19	16.3
Quarries (sand, stone, loam)	2.4	3.9	4.1	3.7	4.4	3.3

Source: Mining Review 2011[3].

Table 3.2: Number of Persons Employed in Mining

Sector	*2009*	*2010*	*2011*
Small and Medium scale Gold and Diamond mining	9,410	10,781[1]	11,672
Large scale – Bauxite mining	1,112	1,073	1,072
Quarries (including sand and loam pits)	267	316	338
Total	10,789	12,170	13,082

Source: Mining Review 2011[3].

mineral processing and beneficiation practices. Indeed, with its own research and development department the local bauxite industry grew a skill pool which after the decline of the sector in early 1990s, was adsorbed into the local gold and diamond sector, as well as other industrial sectors outside of Guyana. Quarry products (sand, stone, and loam) are also extracted, and most of the products are consumed in the local construction industry and for infrastructural development and sea defenses.

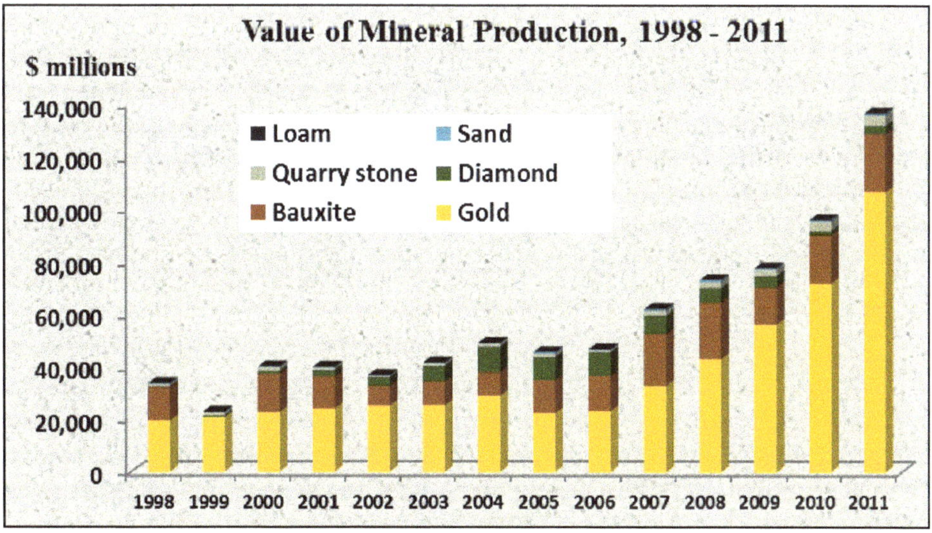

Figure 3.2: Value of Mineral Production.
***Source*: Mining Review 2011[3].**

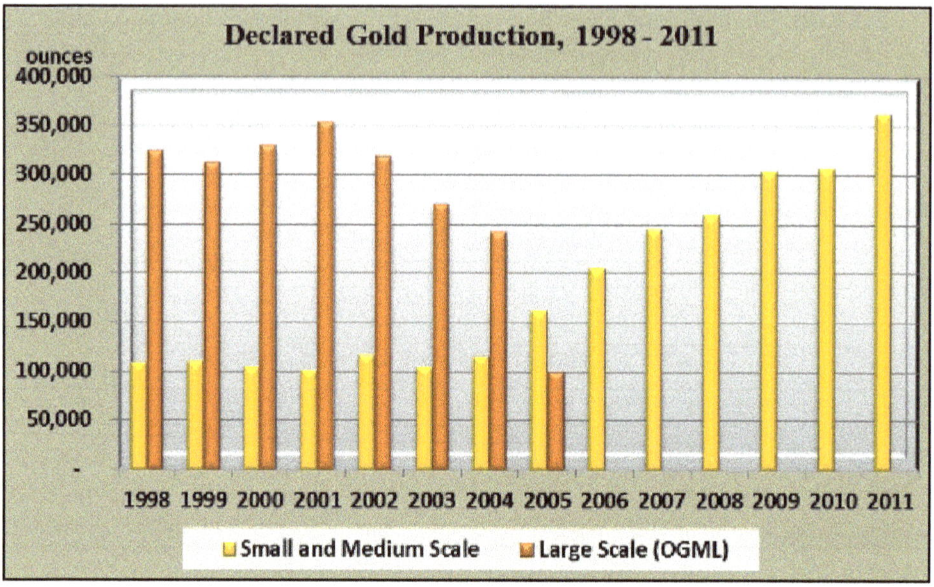

Figure 3.3: Declared Gold Production.
***Source*: Mining Review 2011[3].**

The only large scale gold mining operation at Omai, which was operated first by Cambior Incorporated (1991-2006) then IAMGOLD Corporation (2006-2007) until its closure, maintained mineral processing practices in keeping with Canadian and international standards. This operation peaked in 2001 producing 354,300ounces of gold, processing 21,000 tonnes ore/day, averaging a grade of 1.61gAu/t [4, 5]. The mineral recovery methods in the small and medium scale gold and diamond mining have generally remained the sluice box and batel for gold, and the lavador (Brazilian-Jig) and hand-sieves for diamonds, where recovery is often less than 50 per cent. Therefore, many small scale miners work previously mined land or tailings for a second and third time, thus the grade of the ground is low on the second pass and much lower on the subsequent passes. This decreasing grade of ground is rarely met with improved production efficiency and mineral recovery. These operators also tend to target the rich alluvial flats type material, which are generally unconsolidated and are easier to process with the kind of equipment they possess, but the availability of these is diminishing. Also, the current environmental regulations make it unlawful to mine river banks.

Some miners have taken the initiative and purchased or fabricated screening equipment, impact mills, gravity tables and bowl concentrators and they must be applauded and encouraged, since the GGMC encourages all initiative and improved methods, practices and mineral recovery. However, these efforts occur in less than 50 per cent of current operations. Additionally, hikes in mineral commodity prices on the world market have resulted in greater foreign exchange earnings from gold and diamond mining operations over the last three years, and in 2011 this sector

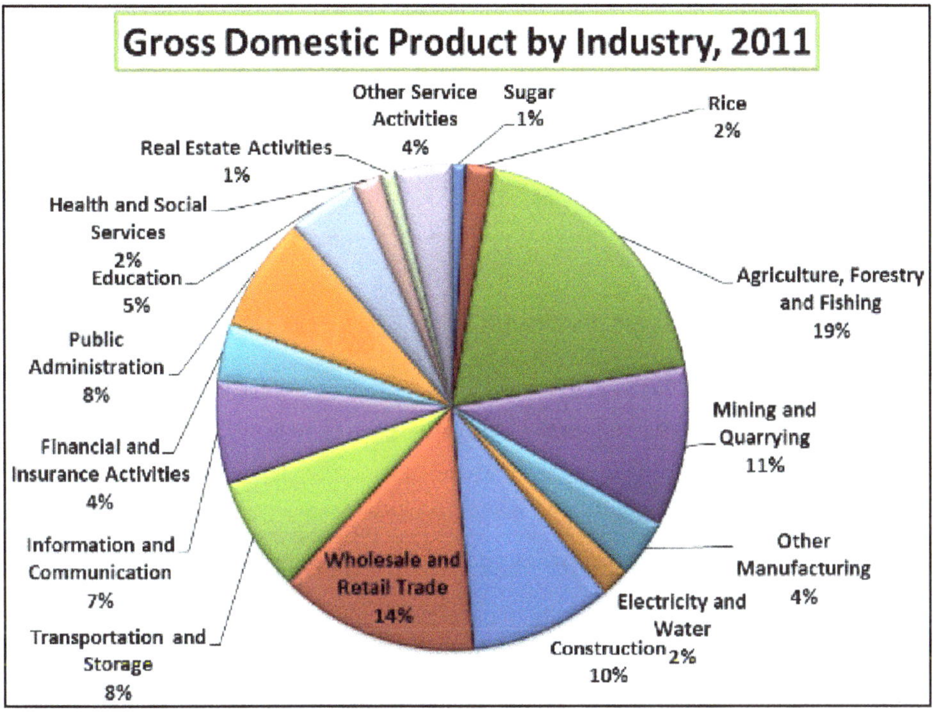

Figure 3.4: Contribution of Mining to Guyana's GDP.
***Source*: Mining Review 2011 [3].**

contributed to over 75 per cent of the Mining Industry's input to Guyana's GDP (the Mining Industry's overall contribution was 11 per cent).

Strong mineral commodity prices have attracted more entrants to the small and medium scale sectors, and while the positive contribution to Guyana's economy is evident, the environmental footprint of these operations has to be strategically addressed. While some mine operators have taken steps to incorporate tailings management as part of production planning and operation, the majority still treat this as an afterthought which is expensive.

With the extra earnings from the current high gold prices miners are encouraged to invest in planning, mineral deposit evaluation and equipment for improved mineral recovery, and to plan and budget for mine reclamation and closure, all of which are promoted or required for mineral exploitation in a Low Carbon Development Strategy (LCDS) environment. The national LCDS sets out Guyana's approach to transition to a green economy. Its stated aim is to combat climate change while simultaneously providing opportunities for economic growth and development. A major objective is to receive payments for avoided deforestation and to invest these payments into sustainable development of its natural resources. The imminent international ban on mercury trade in 2013 increases the urgency for the development of mercury free methods for gold recovery, and this has been accorded the highest priority by the

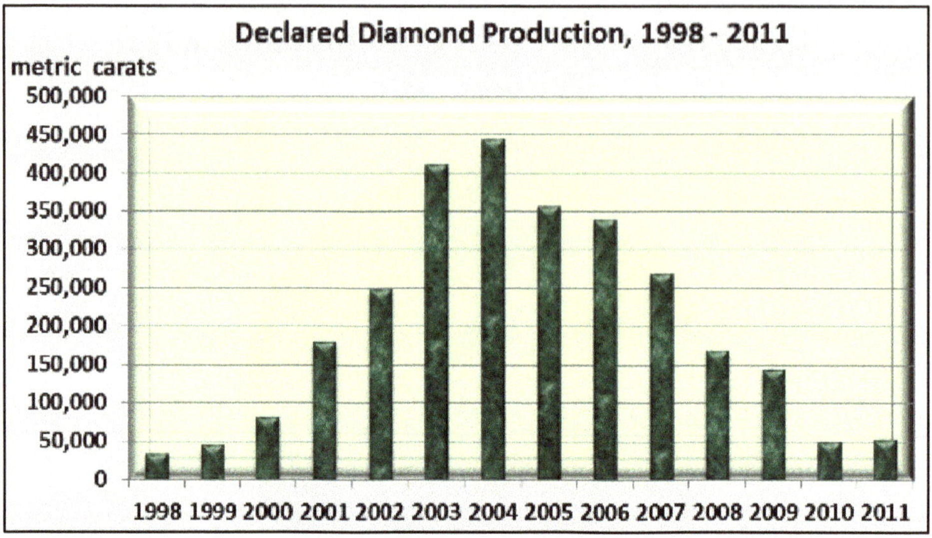

Figure 3.5: Declared Diamond Production.
***Source*: Mining Review 2011 [3].**

GGMC. The opportunity for extracting more mineral products and promoting mineral diversification will also be realised with better planning and evaluation.

2. Mineral Processing Initiatives by the GGMC

Improving mineral recovery has engaged the attention of the Commission from its inception and through workshops and seminars, discussions have helped the small and medium scale sector become more mechanised and improve productivity. Operators have invested in earth- moving equipment, excavators, bulldozers, and slurry pumps giving a boost to these equipment suppliers, but it has not been the same for mineral recovery equipment.

In 1993 the Guyana Geology and Mines Commission, recognizing the importance of small and medium scale miners to the Guyana economy and the difficulty they have in raising capital to develop and evaluate improved gold recovery techniques, embarked on a three (3) phased programmeto design and develop demonstration plants to introduce improved gold recovery methods for land and river based operations. Complementary laboratory facilities to support the demo plants and for mineral processing investigative work, were also planned.

Phase 1 involved finalizing the engineering plant design; generating the equipment list for both the demonstration plant and laboratory; and the procurement of this equipment. Phase 2 involved assembling the plant; installing laboratory equipment; training staff; and testing the plant. Phase 3 involved demonstrations to miners; testing ores; and custom milling. The total estimated cost at that time was US$340,000 and with the financial constraints at that point in time, this did not move beyond the conceptual stage. Nevertheless, this concept was presented at the first biannual Guyana National Mining and Quarrying Conference, hosted by the GGMC in August 1993. At this forum also, the thrust was made to encourage the small and medium scale sector to mine with the environment in mind, by tying improved mineral recovery to improved environmental management.

Over the years the approach was modified. One approach was to improve traditional methods and through the Canadian International Development Agency (CIDA) sponsored Guyana Environmental Capacity Development (GENCAPD) project, work was done to improve sluice box efficiency. Here radiotracers were used to evaluate sluicebox performance, a technique utilised by the Canadian Yukon based consultant Mr. Randy Clarkson. In this method, selected gold particles (^{197}Au) of various sizes, irradiated in a nuclear reactor to ^{198}Au were utilised. The radiation released made the particles (tracers) identifiable with a sctintillometer. This technique [6] enabled backward particle tracking where about 75 irradiated particles are introduced into operating sluicebox. The tracers were in three size ranges of 600, 300 and 150 microns, and each size fraction was introduced separately, with the distribution mapped before introduction of the next size fraction. The sluicebox was operated for a predetermined period then stopped, the distribution of radioactive particles mapped (location tracers are recovered within box). The gold recovered by standard methods, amalgam was digested with nitric acid-HNO_3, and the tracers separated and counted. Use of the tracers enabled the determination of sluice box recovery by grain sizes of gold and aided the demonstration of improved recovery of various size fractions after modifications were made to the sluice box recovery system.

Figure 3.6: Photo: Left–Old Sluicebox; Right–Refitted Sluicebox in Puruni.

Figure 3.7: Photo: Miner Standing Next to New Refitted Sluicebox in Operation in Puruni.

The possibility of using the lavador for gold recovery was examined. The lavador is a jig introduced by Brazilian miners to Guyana for diamond recovery and because local miners had adapted to using this jig.

The lavador is found to be similar to the Yuba type jig, with the moving diaphragm to the side of the jig, instead of the bottom, as is in the case in most jigs. It was felt that if it was successfully modified for recovering gold, it would be easier to have local operators use this equipment that they were already familiar with for gold recovery rather than other equipment that was new to them.

The use of practical demonstrations as a tool to encourage miners' interest in alternative mineral recovery methods has been on going and continues. With assistance from the World Wildlife Fund (WWF), the GGMC successfully completed field demonstrations of the cleangold sluice, the blue-bowl concentrator, and the shaking table. Supporting mine operators willing to utilise improved recovery methods is a part of the GGMC's mandate, and the GGMC took its Knelson concentrator to two mine locations and examined with the operators using the concentrator to upgrade the sluice box concentrate. The use of the Knelson concentrator was also demonstrated in three mining areas at Omai, Mahdia and North West District.

All the foregoing efforts target alluvial and saprolitic material, and as these keep diminishing the more consolidated hard rock material is being processed by small and medium scale miners and leaching as a method for extracting gold is also being introduced. In this effort, GGMC is developing its own capacity in cyanidation technology for gold extraction. Within the last two (2) years GGMC has hired and trained technical staff of the Mineral Processing Unit as trainers in the use of cyanidation technology for gold extraction. Theoretical training on Gold Ore Characterisation and Processing was provided in December 2010, with sessions given by Dr. Marcello Viega, Associate Professor at the Norman B. Keevil Institute of Mining Engineering, University of British Columbia in Canada. Dr. Viega, a prominent member of the international mining community, is also head of the United Nations Global Mercury Project.

The major subject areas covered were Gold Process Mineralogy; Gold Concentration Methods; Cyanide Leaching of gold; and Environmental and Health Implications of Gold Processing. GGMC's Mineral Processing and Mining Engineers, Environmental Officers and Chemists, and Technical Officers of the Guyana Gold and Diamond Miners Association (GGDMA) participated in the course. Cyanide leaching is particularly efficient for the recovery of fine sizes of gold. This was followed by practical training of MPU staff as trainers in the practical aspects of ore concentration and cyanide leaching, at the state of the art Research and Development facility being established at Linden, this year. Then the MPU trainers were evaluated as trainers while training a cross-section of GGMC professional and sub-professional staff. The next aspect to be completed is for both the staff and facility to be audited and certified. This is being done in partnership with CIDA.

The Research and Development facility is being established at Linden, for the practical aspects and applied research into ore concentration and gold recovery by gravity and chemical separation methods that eliminate or greatly reduce mercury use. This facility is intended for:

- Supporting field programmes.
- Building skills and capacity for sample preparation; ore characterization; ore concentration; mineral extraction and recovery.
- Research projects to promote improved mineral recovery and the use of mercury-free gold recovery methods; extraction of available mineral by-products in existing operations.
- Developing Pilot Plant facilities to study mineral recovery units under field-like conditions.

The facility contains several Laboratories for Sample Preparation, Pre-concentration of ores, Cyanide Leach Technology, Sedimentology and Hydrology and General Chemistry. Coupled with the Laboratories, in the same building, a functional Pilot Plant is being developed, which will be outfitted with larger processing units such as jigs, spirals and ore grinding mills.

In 2011, GGMC co-hosted training and technical field demonstrations on the use and benefits of the Knelson MD3 laboratory scale centrifugal concentrator in Georgetown, and at mining areas at Omai, Mahdia, and Purple Heart and Big Creek in the North West District. Miners were encouraged to bring samples of feed to the sluice box and tailings from the sluice box to have them tested for free gold content. In this way, the high efficiency of the Knelson concentrator was demonstrated, and

Figure 3.8: Photo: Randy Clarkson Recovering Lost Tracers in Tailing with Scintillometer.

participating miners were able to see how much gold is being lost by using the Sluice Box, the prevalent method of gold recovery. Demonstrations of the froth flotation as a non-mercury method for recovering gold were also conducted using a laboratory scale mechanical cell.

Technical information on gold recovery methods, in the form of brochures specially prepared by MPU staff, is provided to Miners. The brochures explain the importance of proper equipment selection for recovery of a range of sizes of gold, with emphasis of fine grain sizes and the important parameters to be considered, such as grain size of ore (gravel) that is being processed.

3. Results and Discussions of the Initiatives

From the evaluation of the sluice box with radiotracers, it was confirmed that most of the gold (90 per cent) is recovered in the first 1/3 of the sluice box length. Sluice boxes in excess of 20ft in length were found on operations, some were also too wide. The effective use by miners of expanded metal and angle iron riffles with matting were also some of the benefits of this initiative, and boxes were modified in keeping with results of findings by adjusting the length, angle, and refitting them with angle iron and expanded metal riffles and nomad matting. The limitations of

Figure 3.9: Photos: Radiotracers Flagged and Measured.

this method however meant there was not a complete "buy in" of the technique by local mine operators. The tracer gold morphology was different from the local gold therefore conclusions on recovery of *in situ* (native) gold could be misleading, and the alternative of irradiating gold from each mine location would be too costly. The method also did not address the issue of fine gold recovery. Additionally the technique was proprietary therefore not transferable and could not be a long term solution for the Guyanese miner (Clarkson would always be needed). It must be pointed out that some miners did hire Clarkson at their own cost and reported improvement.

The cleangold sluice was not readily accepted. First there was the issue of sourcing magnetite which was not available in all the areas where the small scale miners being targeted were working. Then when field demos were done the operators argued that it was faster to operate the batel to upgrade gold than to use this technique. There was a similar argument with the blue-bowl concentrator; mine operators felt this was too slow.

Promoting the use of the shaking table and centrifugal concentrators were successful as far as showing the suitability of these devices for fine gold recovery, but for the small miner the argument was these were too costly. Nevertheless, one operator did invest in using a centrifugal concentrator to clean up after sluice box recovery.

Training of staff as trainers in preconcentration and cyanide leaching is a great plus in that GGMC staff are better equipped when in field, to listen to mine operators and give technical assistance. The theoretical training with Dr. M. Veiga, enabled participants to better understand many concepts and gain a better perspective on mineral recovery in general, appreciating that the key lies in understanding the ore that is being processed, its characteristics and how the nature of the ore determines

Figure 3.10: Composite Photo: Mineral Recovery Laboratory at Linden R&D Faculty.

the recovery method, and therefore the equipment to be used. GGMC also now has a facility for further investigation of samples collected from mine operators, and can now assist the sector with information such as the gold distribution in samples, if the gold is gravity recoverable, which gravity method is suitable; if leaching is to be utilised how to go about this technique. Another plus is the drafting of procedural manuals and Standard Operating Procedures (SOPs) which in addition to helping sustain training and skill developed at GGMC can be modified for training and transferring skills to the sector.

With these on-going activities, the targeted results include enhanced mineral recovery; the extraction multiple mineral products; and better tailings management. These activities will foster more collaboration between GGMC and miners, and inputs from miners may include collecting and submitting mineral samples for investigation; sharing the results of miners' investigations; participating in workshops and training programmes; and implementing improved techniques in operations. However, the need for exploration and proper evaluation cannot be over emphasised. This includes the pre-determination of the amount of overburden, type of mineral bearing material, the amount of recoverable minerals; and the grade of the deposit.

While all operators are encouraged to hire the relevant skills and services to help them, it is recognised that those who are better organised and doing well may be able to accomplish this, but this is still costly for the smaller miner and the many new entrants in the local sector. Here the GGMC has to organise and quickly refine its strategy to meet the current demands in this rapidly growing sector. With better evaluation data, more dry stripping and mining with excavators is encouraged since this will facilitate mining and stockpiling, along with better equipment selection and matching for mineral recovery. This will also improve tailings management for small and medium scale operations, because at the production stage the decision would be made by virtue of the type of equipment selected, on the kind of material to be released to the environment, how it would be released, and the quantity. Tailings management would not be an expensive afterthought, but part of the production plan.

The capability to train both the Commission's staff and miners in cyanide recovery of gold is an added plus since this is promoted as a mercury free method for gold recovery, as well as a method for fine gold recovery. The GGMC promotes this technique for treating concentrates and not the whole ore, so that capacity for ore concentration was also developed at the same facility.

Mercury, by law, is only to be used to recover gold in the final concentrate; however there are still operations using mercury in process. These operators still have the belief that if more mercury is in the system, then more fine gold is recovery. Getting across technical information to the miner in way it can be understood and utilised is an on-going issue that is being addressed.

Current operations are mostly geared for single product extraction (gold or diamonds) and no other mineral by-product is considered for extraction. Preliminary tests [7] have shown other mineral products for industrial and commercial uses are available in many operations, but are discarded as waste, and would have to be rehandled at a cost when market opportunities are presented. Mineral samples from dredging operations in Mazaruni and Essequibo rivers have been analysed and showed the presence of ilmenite, zircon, garnet and rutile, all of which have industrial applications [7].

Conclusions–The Way Forward

In concluding it must be acknowledged that the traditional mining by experience and intuition is becoming impossible because of the current need for a systematic approach that considers the environment, and which is efficient and profitable to the investor. Under current regulatory requirements, in order to exist within an acceptable economic scale of function, the Guyana medium and small scale miners must have, or be able to source, the management and technical skills to exploit the mineral resources and not compromise the environment. GGMC's strategy is aimed at providing information gained through applied research, to assist and give guidance to the miner to adopt the systematic approach that is needed for environmentally friendly, efficient and yet profitable operations, in order to ensure the survival of local small and medium scale operations.

The GGMC is working towards the desired result of each small and medium scale miner being able to select mining and mineral recovery methods which would give optimum returns on his or her investment. The focus would be developing simple process options that enable the local miner to get optimum returns on his/her investment. Therefore, using the information generated by applied research mineral processing activities, GGMC will facilitate the economic assessment of mineral recovery options to assist the miner to develop proposals that could be taken to a bank or other serious investors.

None the less, there are challenges going forward, for example, while mercury use for gold recovery is being discouraged internationally, many local operators believe it is still the best option for them. It is cheaper and this is what they are familiar with. Alternative gravity methods and chemical methods have not been embraced by all stakeholders of the small and medium scale sectors, but more and more operators are acknowledging the need for improvement. GGMC, by practice, research, and ongoing training and exposure, has to continue to build capacity for using and optimising alternative mineral processing equipment for improved gold recovery without mercury use and to demonstrate the economic benefits of the alternative processes. GGMC then has to transfer these skills and knowledge through technical assistance to medium and small scale miners and to monitor, assess and give timely support to sustain these changes.

Improved gold recovery methods include chemical methods that are not traditionally used by small and medium scale operators, because of the skills required to maintain the protocols and standards for the safe use of cyanide technology, for instance. Miners and stakeholders will have to be convinced that medium scale operators can be trained and certified to adapt these.

The Guyana Geology and Mines Commission is assiduously working to achieve the above, and is committed to providing the finances; upgrading skills and knowledge of staff; hiring skills that are needed; recruiting and training staff; and encouraging active participation in the industry.

5. Acknowledgements

The author gives grateful thanks to Commissioner, Mrs. Karen Livan for taking the time from her very busy schedule to read and edit this paper. Commissioner (retired), Mr. William Woolford is acknowledged for insights of the Commission's efforts to improve mineral recovery in Guyana prior to author joining the GGMC.

6. References

1. Livan, Karen Editor, 2011: The GGMC Mining Supplement 2011.
2. Guyana National Printers, 2005: The Guyana Official Gazette 19 March, 2005. Legal Supplement B; The Guyana Mining (Amendment) Regulations 2005.
3. Abrams, Wycliffe, 2012: Guyana Mining Summary Review.
4. Nkofi, Kwesi: Enforcement of Compliance Requirements at Omai Gold Mines Limited- Guyana.
   http://www.inece.org/3rdvol 1(accessed Sept 14, 2012).
5. Omai Gold Mines- Source Watch: Extract from Cambior Annual Report 2005, p. 13.
   http://www.sourcewatch.org/(accessed Sept.14, 2012).
6. GGMC/GENCAPD Files: Use of Radio-tracers in the Evaluation of Sluicebox Efficiency in Guyana
7. McDonald Dianne, 2005: Possible Mineral By-products of the Local Gold and Diamond Mining Industry, pp. 5-14
8. McDonald Dianne, 2009: Small and Medium Scale Mining in Guyana- Issues to be addressed to Improve Mineral Recovery and Increase Profitability.

Chapter 4

An Application for Java Natural Silicates Mineral and Recycled Amorphous Silica and its Beneficiation for Production of Advanced Material and Pharmaceutical Raw Materials

*Eko Tri Sumarnadi Agustinus[1], Happy Sembiring[1], Mutia Dewi Yuniati[1], Effendi[2] and Anggoro Tri Mursito[1]**

[1]Research Centre for Geotechnology, [2]Research Centre for Chemistry, Indonesian Institute of Sciences (LIPI), Jl. Sangkuriang, Gd. 70, Bandung 40135, Indonesia

**E-mail: anggoro@geotek.lipi.go.id*

ABSTRACT

The utilization of natural silicates mineral in West Java are significant to increase its added value and has its benefit for securing the demand of advanced material and pharmaceuticals in Indonesia. The mining of industrial minerals, community development and research and development are the bridge for an application of production of advanced material and pharmaceuticals. Two groups of research team have been working for decades to solve and provide options to have a simple and a low cost of mineral processing. The minerals which have been chosen are perlite, bentonite and zeolite. Those resources are enourmous in large areas surrounding the coastal line in Tasikmalaya- West Java. Perlite and bentonite have been used as a raw advanced material for preserving the microorganism/

bacteria called *Bacillus licheniformis (L.P)* and used for reducing the phenols and other organic compounds concentration in waste water treatment and has been applied, particularly, at oil refineries and sugar cane industries. The products are formed in preserved microorganism tablet. Fortunately, the products and processes involved were listed and registered for patent at the beginning of this year.

Introduction

Organic liquid waste which are produced by industrial complex in West Java especially, Indonesia were increased. In 2009, there were about 201,953 industrial units and about 85 per cent of them are producing liquid waste containing organic and then treated at biological waste water treatment especially by microorganism. Microorganism supply and packaging could be found in a liquid culture and powder. Unfortunately, both of products of microorganism packaging are not properly suggested recently due to short preservation of liquid culture and could be infected in the respiration respectively.

The objective of the research were to utilize and demonstrate that natural silicate minerals and recycled amorphous silica that could potentially be used for the preservation of microorganism prior treated by the activation and hydrothermal oxidation.

Materials and Methods/Experimental

Raw natural silicates mineral (bentonite) and recycled amorphous Silica samples were obtained from industrial mineral mining at Tasikmalaya, West Java and geothermal site at Dieng, Central Java Indonesia repectively.

Activation and Intercalation

The activation and oxidation experiments were conducted in a modified adjustable tube furnace reactor (Thermolyne) of 60 mm diameter and 320 mm length equipped with automatic temperature control (Figure 4.1.). Steam was produced by a modified double hotplates (Thermolyne) and borosilicate flat bottom glass flasks (Pyrex) and capable of producing a maximum steam rate at 400 mL/h in the atmospheric pressure. Steam was further heated in a heater coil of stainless steel pipe of 45 mm diameter and 400 mm length. 50g of sample was placed inside the middle of tube furnace reactor, and inlet and outlet of this tube was sealed with glass-asbestos wool. The inlet of the stainless steel pipe chamber was connected into the steam producer and outlet connected to the flash tank with NaOH 0.1 N solution. Next, the sample was heated in an atmospheric pressure while the reaction temperature was automatically adjusted at 800°C at an average heating rate of 6°C/min. The superheated steam was flowed when the temperature reached to 400°C. After the desired reaction time of 60 min, the reactor was rapidly cooled.

The intercalation experiments were only conducted for bentonite with 100 mesh particle size by adding the 10 ppm of EDA (ethylene diamine) at room temperature on agitation plate at 150 rpm for 20 min. Optimation pH, agitation time and rotation and temperature were also conducted at different value.

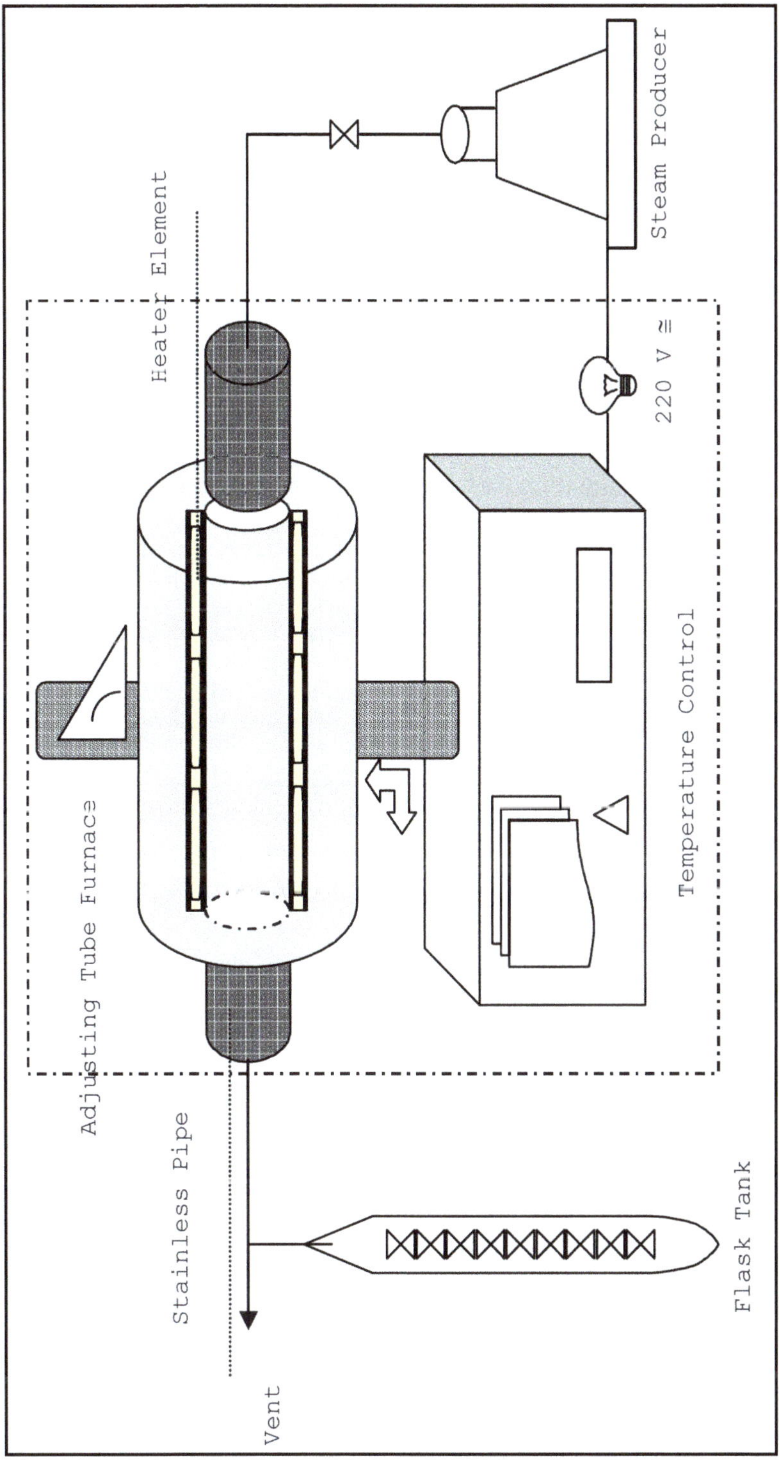

Figure 4.1: Schematic Figures of the Steam Activation Treatment.

Hydrothermal Treatment

The experiments were conducted in a 1 L batch-type reactor (Jaeger Tool) that was equipped with an automatic temperature controller and had a maximum pressure of 40 MPa and a maximum temperature of 400°C (Figure 4.1). The slurry samples were introduced to the reactor without any pretreatment. The amount of the slurry added to the reactor was 700 g, corresponding to 40 g of sample. Next, the raw peat was heated with agitation at 200 rpm while the reaction temperature was automatically adjusted from 200 °C at an average heating rate of 6.6 °C/min. After the desired reaction time of 30 min, the reactor was cooled immediately. After cooling down, the solid and liquid phases were then collected from the reactor and separated by filtration using a water aspirator.

Bacteria Culturing and Adsorption

Several experiments were conducted for preparation of bacteria cultures of *Bacillus licheniformis* such as nutrition media preparation and innoculation. Nutrition media preparation by adding amylum starch with/without malt extract, NaCl 0.85 per cent; NaOH 1N and HCl 1N at pH 5 (without malt extract) and pH 7 (with malt extract) and retention time for 30 min. with the concentration of amylum such as 1.5; 2.0 and 2.5 (without malt extract) and 6.5; 7.0 and 7.5 (with malt extract), while the given concentration for other supporting culture of bacteria such as NH_4Cl; Na_2HPO_4; $MgSO_4$; $FeCl_3$; $CuSO_4$.

Table 4.1: Amylum Starch Preparation and Parameter Involved without Malt Extract

Sl.No.	*Nutrition*	*pH*	*Retention Time (min.)*	*Concentration (g/L)*		
1.	Amilum	5.0	30	1.5	2.0	2.5
2.	NH_4Cl	5.0	30	1.0	1.0	1.0
3.	Na Posphate	5.0	30	1.5	1.5	1.5
4.	$MgSO_4$	5.0	30	1.0	1.0	1.0
5.	$FeCl_3$	5.0	30	0.1	0.1	0.1
6.	$CuSO_4$	5.0	30	0.5	0.5	0.5

Table 4.2: Amylum Starch Preparation and Parameter Involved with Malt Extract

Sl.No.	*Nutrition*	*pH*	*Retention Time (min.)*	*Concentration (g/L)*		
1.	Amilum	7.0	30	6.5	7.0	7.5
2.	Urea	7.0	30	2.0	2.0	2.0
3.	Na_2HPO_4	7.0	30	6.0	6.0	6.0
4.	Mg SO_4	7.0	30	4.0	4.0	2.0
5.	$FeCl_3$	7.0	30	0.4	0.4	0.4
6.	$CuSO_4$	7.0	30	2.0	2.0	2.0
7.	Malt Extract	7.0	30	6.5	7.0	7.5

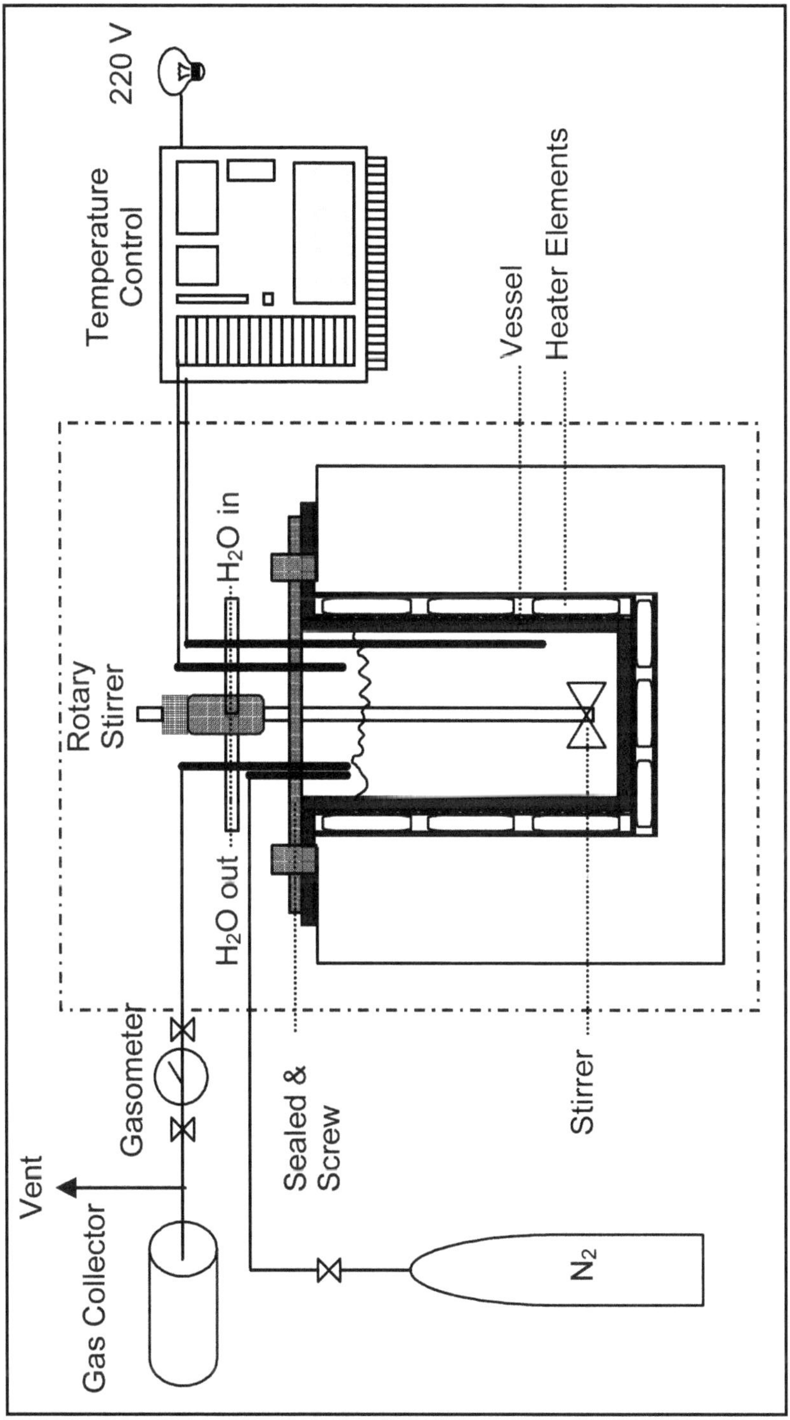

Figure 4.2: Schematic Figures of Hydrothermal Batch Type Reactors.

Bacterial inoculation in amylum starch media such as: bacteria mixing with NaCl 0.85 per cent ± 10 mL in tube and sediment material will be mix in the amylum starch media suspension and then agitation and aeration will be introduced for several days (1 day to 5 days) in 5 L beaker at pH 7.0. Several condition of nutrient will be added by adding 1 L aquadest and amylum starch and the pH will be adjust by NaOH 1 N/HCl 1N. Viability, pH and COD will be measured everyday.

Table 4.3: Sampling Time and Parameter Involved with/without Malt Extract

Sl.No.	*Sample Code*	*Culturing Time (hours)*	*pH*
1	Am 1*)	24	7.0
2	Am 2	48	7.0
3	Am 3	72	7.0
4	Am 4	96	7.0
5	Am 5	120	7.0
6	Am 6	144	7.0
7	AmME1**)	24	7.0
8	AmME1	48	7.0
9	AmME1	72	7.0
10	AmME1	96	7.0
11	AmME1	120	7.0
12	AmME1	144	7.0

Amylum starch media with*)/without Malt Extract**).

At 5 days, the suspension and nutrient adding will be stop, and then aeration will also be stop, and the sediment will be collected for every 300 mL and then mixed with activated minerals in the bioreactor tank with several parameter pH, temperature, electric conductivity, COD and agitation rotation.

Table 4.4: Bacteria Adsorption and Parameter Involved

Sample No.	*pH*	*Temperature(ºC)*	*Agitation (rpm)*
1	6.0	28	250
2	6.5	29	250
3	7.0	30	250
4	7.5	31	250
5	8.0	32	250

Pelletization

Adsorbed bacteria in activated bentonite and amorphous Si will be milled at -100+200 mesh fraction and dried in a oven at 30-70°C for 36-120 hours. Pelletization in a form of tablet of activated bentonite and amorphous Si were done at compressive pressure of 100; 150 and 200 kg/cm^2 with the ratio of material, CMC (Carboxyl Methyl

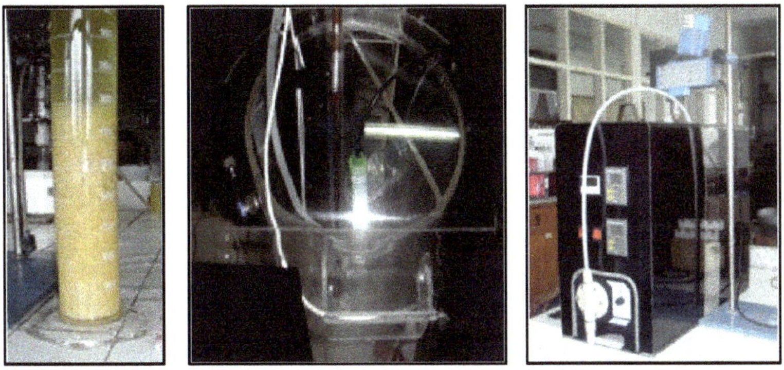

Figure 4.3: Adsorption Tank Picture.

Figure 4.4: Pelletization in a Form of Tablet.

Cellulose), starch and stearic acid of 10: 0.5-2.5: 0.5-2.5: 0.5 at 3-5 mm diameter, 8 mm height and density of 1.2-1.7.

Results and Discussion

Table 4.5 shows the results of the major element of raw bentonite material and recycle amorphous silica measured by Atomic Absorption Spectroscopy (AAS). SiO_2 is the main components followed by Al_2O_3 which are rich in Si and Al, respectively. Other elements in bentonite such as Fe, Na, Mg, K, Ti and Ca are the minor elements and must be treated and reduced in order to obtain and increase the purity of montmorilonite mineral.

The main mineral of bentonite which are analysed by XRD is montmorilonite (80 per cent) and has a formula $M_x(Al^{4-}xMg_X)Si_8O_{20}$ $(OH)_4$. nH_2O (Figure 4.5.). Montmorilonite structure contained for 3 layer of 1 alumina layer (AlO_6) which has octahedral form and enclosed with 2 silicate layer (SiO_4) which has tetrahedral form. In between of those layer of octahedral and tetrahedral, there are monovalent cation and bivalent which are Na^+ = 56.600 ppm; Mg^{2+} = 1300 ppm and no occurence of Ca^{2+} and has a d-spacing for about 14.32 Å. Fig. 6 shows the XRD spectrum of the geothermal waste which obtained from Dieng geothermal plant in Central Java, Indonesia and its consisted of amorphous silica.

Table 4.5: The Amounts of Major Elements of Raw Material

Oxide (Major elements)	*Concentration (wt. per cent)*		
	Bentonite (Ca)	*Bentonite (Na)*	*Amorphous Silica*
SiO_2	51.79	53.49	84.2
TiO_2	0.69	0.65	n.a
Al_2O_3	11.44	11.32	0.14
Fe_2O_3	4.41	3.21	n.a
MnO	0.78	0.77	n.a
MgO	0.57	0.55	0.039
CaO	0.15	0.05	n.a
Na_2O	7.31	8.35	n.a
K_2O	1.65	1.76	n.a
P_2O_5	0.99	1.32	n.a
LOI	19.01	19.53	n.a

Intercalation

Bentonite layers are naturally aglomerated due to the adhesion force in particle whenever contacted with water. EDA intercalation is one of the method to reduce the adhesion force and to increase distance of layer in montmorilonite structure (d-spacing) due to cationic and organic exhange. Optimalization of intercalation could

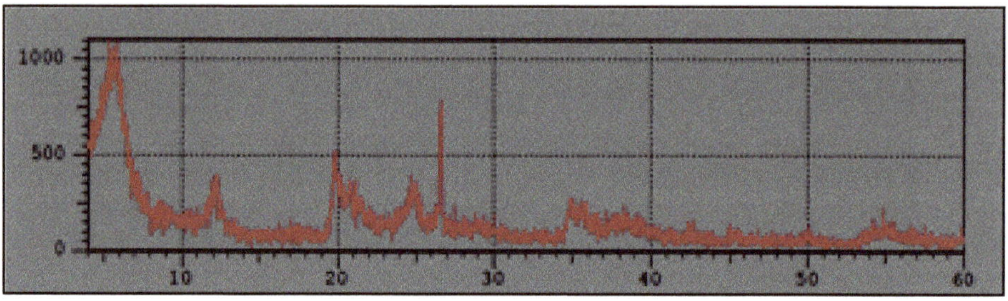

Figure 4.5: Result of X-RD Spectrum of Bentonite.

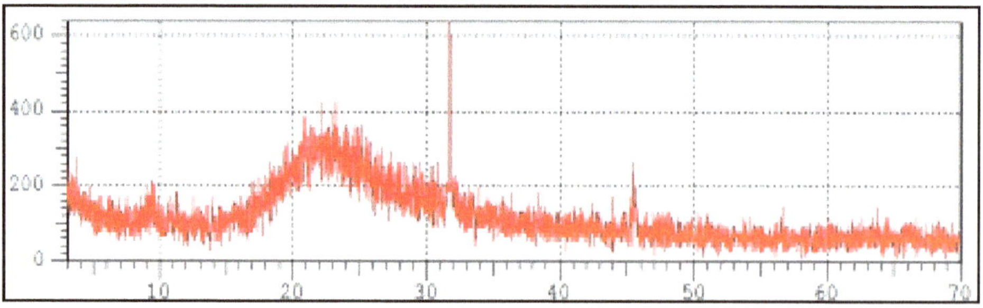

Figure 4.6: Result of X-RD Spectrum of Amorphous Si.

be done by adjusting the experimental parameter such as pH, residence time, EDA concentration, fraction size, temperature and agitation speed.

Intercalation process and its optimal value on EDA adsorption have been done at pH 3, 4, 5, 6, 7 and 8 and also measured for cation exchange capacity (CEC) on Ca, Mg and Na. At an optimal pH 5 and the adsorption of EDA at 67.21 per cent results on reducing of Na-CEC for 99.54 per cent.

At intercalation process experiment at optimal pH 5 and residence time at 20, 40, 60, 80, 100 and 120 min. results an optimal residence time at 60 min. and the adsorption of EDA at 71.25 per cent results on reducing of Na-CEC for 99.42 per cent. Other optimizing experiments on EDA adsorption in activated bentonite results at 20 ppm and the adsorption of EDA at 80.5 per cent results on reducing of Na-CEC for 99.16 per cent. At the optimum temperature of 32°C results the adsorption of EDA at 83.7 per cent and resulting on reducing of Na-CEC for 99.44 per cent. Also, at the optimum particle size of 200 mesh results the adsorption of EDA at 89.41 per cent and resulting on reducing of Na-CEC for 99.08 per cent.

Bacteria Viability

A 5 days mixing of bacteria and the media which is rich in nutrition and then followed by suspension sampling every 24 hours and measure for viability which will know the cell growth (cell/L). Table 4.6 explained the viability results on the bacteria culturing. Increasing the viability was observed in different amylum extracts. In malt extract, results the higher viability than without malt extract.

Table 4.5: Viability of Bacteria

No.	Sample Code	Culturing Time (Hours)	Viability (cell/L)
1	Am 1*)	24	7.11E + 9
2	Am 2	48	4.32E +10
3	Am 3	72	5.01E + 11
4	Am 4	96	9.17E +12
5	Am 5	120	1.02E + 15
6	Am 6	144	1.02E +15
7	AmME1**)	24	7.21E + 9
8	AmME1	48	4.52E +10
9	AmME1	72	5.17E + 11
10	AmME1	96	9.31E +12
11	AmME1	120	1.08E + 15
12	AmME1	144	1.08E +15

Bacteria Adsorption

Table 4.6 shows the result of bacteria adsorption in the bioreactor mixed with the activated and intercalated materials results in a different number of adsorbed bacteria. By a combination between the acid washing and steam activation of materials will result an increasing the number of adsorbed bacteria. The minimum requirement of the amounts of adsorbed bacteria is 1.6. 10^6 cell/g.

Table 4.6: Bacteria Adsorption at Selected Materials

No.	*Sample Code*	*Bacteria Number (cell/g)*
1	Steam activation 900% C, 1 h.	4. 10^5
2	Hydrothermal Treatment 300% C, 1 h, 350 rpm	5.2. 10^6
3	Acid washing HCl 1 M at 110% C, 4 h, Steam activation 900% C, 1 h.	6.2. 10^6
4	Acid washing HCl 2 M at 110% C, 4 h, Steam activation 900% C, 1 h.	5.6. 10^6
5	Perlite, raw materials, intercalated	4. 10^6

Pelletization

Physical characteristic of pellet in a form of tablet such as density, hardness and dissolubility was evaluated, observed and explained for phenomenon occurred. As an additional material such as CMC (as bonding material), starch (depressant agent) and stearic acid (slicker) will be need as a major issue on pelletization.

Megascopic observation results shows that compactness is roughly smooth and some physical defect on ring cap due to the symbol form. Table 4.7 shows that the optimal tablet form and physical properties which were proposed are density 1.75, hardness 13.55 kg/cm^2 and dissolubility 100 per cent in 1.67 min./100 sec. at the compressive strength on 200 kg/cm^2 for the activated bentonite 76.9 per cent; CMC 15.4 per cent; Starch 3.9 per cent and stearic acid 3.8 per cent. Table 4.1 shows that the influence of compressive pressure is more dominant than the amount of bonding

Figure 4.7: The Prototypes Picture of Preserved Microorganism Tablet.

Table 4.7: The Results of Pelletization and Parameter Involved

No.	Comp. Pressure (kg/cm²)	Bentonite	CMC	Starch	Ste. Acid	Density (g/cm³)	Hardness kg/cm²	Dissolubility (sec.)
A1	100	10	0.50	0.50	0.50	1.20	6.50	30
A2	100	10	1.00	0.50	0.50	1.25	7.00	35
A3	100	10	1.50	0.50	0.50	1.20	7.50	35
A4	100	10	2.00	0.50	0.50	1.25	7.50	35
A5	100	10	2.50	0.50	0.50	1.20	7.50	45
A6	100	10	1.50	1.00	0.50	1.25	7.55	40
A7	100	10	1.50	1.50	0.50	1.30	7.50	40
A8	100	10	1.50	2.00	0.50	1.35	7.55	40
A9	100	10	1.50	2.50	0.50	1.30	7.55	50
B1	150	10	0.50	0.50	0.50	1.30	7.60	40
B2	150	10	1.00	0.50	0.50	1.30	7.55	40
B3	150	10	1.50	0.50	0.50	1.35	7.60	40
B4	150	10	2.00	0.50	0.50	1.30	7.60	45
B5	150	10	2.50	0.50	0.50	1.45	7.60	50
B6	150	10	1.50	1.00	0.50	1.10	8.00	55
B7	150	10	1.50	1.50	0.50	1.50	8.50	65
B8	150	10	1.50	2.00	0.50	1.50	8.80	60
B9	150	10	1.50	2.50	0.50	1.65	10.00	70
C1	200	10	0.50	0.50	0.50	1.60	11.50	85
C2	200	10	1.00	0.50	0.50	1.60	12.00	90
C3	200	10	1.50	0.50	0.50	1.70	13.00	95
C4	200	10	2.00	0.50	0.50	1.75	13.55	100

materials. At compressive pressure for 100 kg/cm² and 200 kg/cm² results to have different value of hardness at 7.5–13.5 kg/cm². On the other hand, increasing the amount of bonding material at 0.5–2.5 g result to have different value of hardness at 0.5–2 kg/cm². At compressive pressure for 100 kg/cm² and 200 kg/cm² results to have a density of tablet A4, B4 and C4 at 1.25–1.75 g/cm³. These results indicates that the increasing of compressive pressure were not equal to the increasing of density of tablet, due to the CMC which have the spring back properties (James Reed, 1998). Porosity of a material has also the important aspect on the formation of density, form of tablet and also the particle size fraction (Mc.Geary, 1998). The hardness of tablet 13.55 kg/cm² is on the standard range for 10–20 kg/cm².

Conclusion

In this research, a study and an application of activation and hydrothermal oxidation of raw bentonite, perlite and amorphous silica at different condition have been conducted. All treated material can be potentially used for the preserved microorganism material. The preserved microorganism tablet has been applied for the waste water treatment in industrial scale and the process will be implemented in a pilot scale project in 2014.

Acknowledgement

The authors gratefully acknowledge the contributions of PT. Geodipa Energi, Indonesia for providing us the sample and serving us during the field research and experiment and contribution of the Laboratory of Mineral Processing, Beneficiation and Recycling – Research Centre for Geotechnology LIPI for providing us the equipments and analytical instruments. Financial support was partly provided by a research grant of Kompetitif LIPI 2012-2013 in the advance material and nano technology programme.

References

1. Agustinus ETS, 2009, Sintesis Hidrotermal Atapulgit Berbasis Batuan Gelas Volkanik (Perlit) : Perbedaan Perlakuan Statis dan Dinamis Pengaruhnya terhadap Kualitas dan Kuantitas Kristal, Prosiding Seminar Material Metalurgi, Penguatan Riset Material dan Metalurgi Menuju Industri yang Ramah Lingkungan, Pusat Penelitian Metalurgi-LIPI, Kawasan PUSPITEK Serpong-Tangerang.
2. Balbich, H and H.A, Davis, 1981, Phenol: A Review of Environmental and Health Risk Regulat : *Toxicol Pharmacol* 1: 90-190
3. Bertone.J.F, Cicevon J, Wahi K.J, Boswarth J.K and Calvin V.L, 2003, Hydrothermal synthesis of Quartz Nanocrystals, *Nano Letters*, Vol 3, No 5, p: 655-659.
4. Hirnawan F, 2007, RISET, bergulirlah PROSES ILMIAH, Unpad Press, Bandung.
5. Morey. G.W, 1953, Hydrothermal Synthesis, *Journal American Society*, 36. 279.
6. Sembiring,T. and A. Susilorukmi, 1997, Pengujian Pellet Mikroba Penghancur Phenol, *Proceeding Seminar Nasional Fisika Terapan dan Lingkungan* '97, Serpong 1997, ISBN 979-8580-15-X;979-8580-17-6
7. Sembiring,T and H. Sembiring, 1997, Pelletisasi Mikroorganisme Bakteri Penghancur Phenol : Buatan Pellet dengan Bahan Tambahan Mineral Alam. Prosiding Seminar Nasional Biologi XV., Lampung, ISBN : 979-8287-17-7
8. Sembiring,T and H.Sembiring, 1998, Optimalisasi Sifat Campuran Zeolit dan Dolomit sebagai Media dalam Pelletisasi Bakteri, Buletin IPT
9. Sembiring T, A. Susilorukmi and H. Sembiring, 1998, Uji Viabilitas Bakteri Penghancur Phenol dalam Pellet Dolomit, Zeolit dan Karbon Aktif. Teknologi Indonesia.
10. Somiya S, and Roy R, 2000, Hydrothermal Synthesis of Fine Oxide Powders, *Bull. Mater Sci.*, Vol 23 No.6, p : 453-460.
11. Sudjana, 1980, Disain dan Analisis Eksperimen, penerbit TARSITO, Bandung
12. Sudradjat, A, 1999, Teknologi and Manajemen Sumberdaya Mineral, ISBN 979-8591-99-2, Penerbit ITB Bandung.

Chapter 5

The Minerals of Jordanian Oil Shale: Characterization and Benificiation

Awni Y. Al-Otoom

Chemical Engineering Department, Jordan University of Science and Technology, Irbid, 22110, Jordan
E-mail: awni_otoom@just.edu.jo

ABSTRACT

Oil shale in Jordan is an important and unutilized energy source in the country. More than 50 billion tons of oil shale deposits are discovered in different areas especially in the middle and southern parts of the country.

A characterization study was performed on minerals from oil shale especially those obtained from combustion of Jordanian oil shale. This characterization utilized different analytical techniques. These include scanning electron microscope with energy dispersive spectrum analysis, X-ray diffraction, and Qemscan. The dominant mineral phase found in the oil shale was the calcite. Different proportions of clays, quartz, and phosphate were also present. There was no consistent distribution of minerals in oil shale particles. However, some pure calcite particles were also found in the sample of oil shale. Sulfur was found to be mainly in the organic part of the oil shale. The organic part was found to be encapsulated by a shell of mineral, mainly calcite. Some inorganic sulfur was detected only by the Qemscan. During combustion, it is expected that the oxidation product of organic sulfur react with the calcite, which acts as a sulfur sorbent media.

The floatation of oil shale has shown a poor recovery of oil rich components with a maximum of 3 per cent enhancement when using fuel oil #2 as a collector. This poor recovery is believed to be due to fact that kerogen is encapsulated in layers of minerals and that kerogen, during crushing and milling wets most of the oil shale particles making it difficult to adjust the surface properties.

Demineralization by treating oil shale has demonstrated a technical viability and feasibility by increasing the oil yield by approximately two folds. Acetic acid was the most efficient acid in the demineralization process as a result of reacting with most of the calcite forming calcium acetate.

Keywords: *Oil shale, Jordan, Minerals, Beneficiation.*

Introduction

Soaring prices of oil have paved the road to exploring potential unexploited sources of energy. Among these sources is the oil shale, which is considered to be one of the largest sources of hydrocarbon in the world. The oil equivalent of oil shale around the world is estimated to be around 30 times the reserve of the crude oil.

Jordan is considered among the top five countries in the world in terms of the oil shale reserves. Deposits have been found in several places in Jordan, particularly in the central region (Hamarneh *et al.*, 2006). Work is currently undergoing to study the feasibility of producing oil from oil shale. Utilization of this source will assist the economy and provide an alternative source of the demand of energy in Jordan.

Oil shale in Jordan and elsewhere faces many obstacles that hinder its utilization. Among these are the high amounts of inorganic materials, mainly carbonates, which could reach values of greater than 55 per cent of its content. The high sulfur content in oil shale is considered one of the biggest challenges in the utilization of Jordanian oil shale (Al-Harahsheh *et al.*, 2005). Utilization of this source would provide an alternative source to cover the demand of energy in Jordan for decades.

The organic part (kerogen and bitumen) of Jordanian oil shale is estimated to be 22 per cent and the average Fischer assay yield of liquid oil fraction is estimated to be around 13 per cent. (Hamarneh *et al.*, 2006, Dabbas, M. A, 1997, Khraisha, Y., 2000). The inorganic part consists mainly of calcium and magnesium carbonate in addition to quartz, alumina-silicate, pyrite and other trace minerals. The major components and properties of El-lajjun oil shale deposit are shown in Tables 5.1 and 5.2 (Hamarneh *et al.*, 2006).

Beneficiation of organic matter can be obtained via physical or chemical methods. Physical beneficiation methods are based either on the difference in specific gravity or on the wettability difference between organic and inorganic components by floatation techniques. These methods are used for finely ground samples (Vandenbroucke, M. and C. Largeau, 2007). Jordanian oil shale has shown a poor floatability due to the fact that bituminous materials wet all particles during crushing and preparation stages.

This work includes a characterization study for mineral composition of oil shale, as well as some methods of beneficiation including floatation, and demineralization.

Experimental

Characterization of Minerals in Oil Shale

A standard sample of oil shale was brought from the El-lujjun deposit, which is considered the largest deposit in Jordan, by the Jordanian Natural Resources Authority.

Table 5.1: The Average Physical and Chemical Characteristics of the Sub-units in El-lajjun Oil Shale Deposit (Hamarneh *et al.*, 2006)

Sub-unit	*Oil %*	*Moisture %*	*S %*	*C-org %*	*Calorific Val. kJ/kg*	*$CaCO_3$ %*	*Al_2O_3 %*	*P_2O_5 %*	*SiO_2 %*
G	4.3	3.6	1.9	3.6	2580	73.9	2.5	0.5	5.6
F	3.4	2.3	1.7	5.6	2050	71.1	3.1	1.3	6.7
E	5.7	2.8	2.6	8.6	4180	56	4.8-6.7	1.73	12.4
D	11.1	3.4	3.3	12.8	6850	45	5.0-6.7	3.35	13.3
C	7.8	2.2	2.3	9.8	4700	53	4.2-4.8	3.3	11.5
B	12.3	3.1	3.2	14	8100	33	2.6-2.9	3.6	26.1
AL	5.2	2.3	1.7	6.2	4090	60	20.1-6.6	3	22
A	12.6	1.9	3.5	14.5	8170	46	3.4-3.5	2.7	14.6
P	6.1	1.8	1.9	8.3	4200	45	1.1-1.4	9.7	9.1

Table 5.2: Composition of a Representative Samples of El-lajjun Oil Shale (Hamarneh *et al.*, 2006)

Fischer Assay	*Rich*	*Lean*	*Average*
Oil per cent	16.7	8	13.4
Water per cent	2.1	3.3	2.6
Spent Shale per cent	83	76.5	80.3
Gas and Loss per cent	5.7	3.7	4.6
S.G. of Oil	0.977	0.964	0.966

Laboratory ashes from this oil shale were prepared by initially crushing the oil shale to <150 μm, and ashing in an electrically-heated mufe furnace in accordance with the Australian Standards for ashing hard coal. One gram of oil shale is placed inside a silica crucible (diameter is 5 cm) and heated in air to 500 °C in 30 min. The sample is then heated from 500 to 815 °C in 60 min, and kept at 815 °C for 3 h to insure a constant remaining mass and hence complete combustion. XRD analysis was performed on the resulting ash samples.

The characterization was performed using Scanning electron microscope (SEM). SEM is a qualitative or semi quantitative technique useful for obtaining particle composition and morphology. A JEOL JSM-840 SEM with an attached Oxford Isis Energy Dispersive System (EDS) is used in this study. The typical operating parameters for the instrument are: a beam current equal to 3 nA, and an acceleration voltage of 15 kV. The working distance varied from 15 to 39 mm. Beside the chemical analysis, SEM is used to obtain images for the distribution of minerals in oil shale.

Another method of characterization was performed using QEMSCAN analysis. The QEMSCAN system was developed by CSIRO/Australia. QEMSCAN is a relatively new automated system that rapidly determines mineral distributions in ores and processing-plant products. Originally designed for mining and mineral applications, QEMSCAN has also found broad application in the exploration, coal and oil and gas industries (Intellection Co.-Australia Qemscan).

Floatation of Oil Shale

Denver DR floatation machine was used to perform froth floatation, which is an example of a typical froth floatation unit used in the mining industry. A schematic diagram is presented in Figure 5.1, The pulp is introduced through a feed box and is distributed over the entire width of the cell. The pulp comes into contact with the impeller, where it is subjected to intense agitation and aeration. The compressed air enters the cell by impeller rotation.

MIBC (methyl isobutyl carbonel), a widely used frothing agent, was used as a frothing agent. Different floatation agents (collectors were used); these are mainly Kerosen, DMSO, calcium lignosulfonate, sodium lignosulfonate, sulphonated naphthalene, diethyl ketone, and Fuel oil #2.

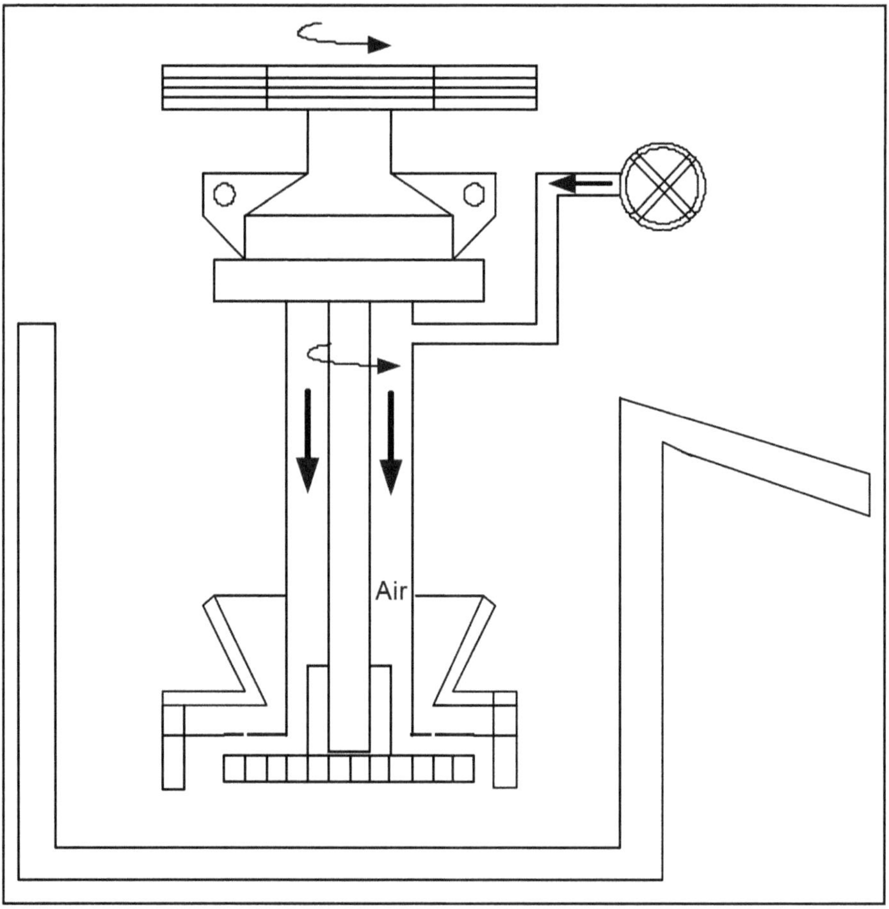

Figure 5.1: Schematic Diagram of the Floatation Cell.

A Sample of oil shale is weighed (150 g) and placed in a beaker which contains the solution of the solvent (water), collectors and frother. Samples were kept for conditioning for a period of 10 minutes. The total volume of the solution is 1 litre.

The solution is then placed in the floatation cell and air flow rate is adjusted to 2 L/min, and the rotation speed of the impeller is kept at 1000 Rev/min. Frother is continuously collected from the cell into a beaker for a period of 30 minutes. Tailing and froth is then collected, washed with distilled water, vacuum filtered and then dried in an oven at 105°C.

Demineralization of Oil Shale

A 5 g ground oil shale (125-250μm) sample was mixed with 20 ml of acid of concentration (5, 10, 20 and 35 per cent). Different acids of different concentrations were used in this study namely HCl, HF, HNO_3, and CH_3COOH. Mixing was performed, while heating using a hot plate, using a low speed mixer until boiling was reached. The mixture was then filtered using vacuum suction through 0.45 μm filter paper and then washed with excess amount of water to remove acid residues. The solid residue was then dried at 60 °C for 24 hours to remove moisture. A mixture of acids was also used (35 per cent HCl, 10 per cent HNO_3, 10 per cent HF) in order to maximize the yield obtained after acid digestion. 50 g of oil shale concentrate was prepared for Fischer assay analysis according ISO 647-74.

Results and Discussion

Characterization of Oil Shale

XRD was performed on combusted oil shale samples. It is clear from Figure 5.2 that calcite is the dominant phase in the minerals of the oil shale. Clays and quartz in addition to the phosphate were also present in different proportions.

The analysis from the energy dispersive spectrum (EDS) in the scanning electron microscope shows that the major phases in oil shale are present as calcite, clay, and quartz. It was surprising to see the oil or kerogen in the oil shale was encapsulated inside a shell of minerals, mainly calcite. The analysis of this kerogen using EDS showed that it mainly consists of carbon, hydrogen and sulfur. Figure 5.3 shows samples of some SEM images for cross-sections of oil shale particles. These images show the minerals and spherical shells which encapsulate the kerogen. A thorough analysis by the EDS–SEM indicated that there was very little inorganic sulfur, which is mainly bounded to iron atoms forming pyrites. Sulfur was mainly present in the organic part of oil shale (in the kerogen).

The Qemscan was also employed to analyze and visualize the distribution of minerals in oil shale. The dominant phase recognized by the Qemscan was also the calcite, which is also represented by the calcareous matrix in Figure 5.4. It is important here to mention that Qemscan detected a considerable amount of pyrite. This was expected since the size of the analysis beam in the Qemscan is 2 mm. Detecting any of the sulfur in the organic part along with iron in the surrounding shell could result in the anticipation of the formation of pyrite. It is worth mentioning that there was no uniform distribution of any of the detected minerals inside any oil shale particle.

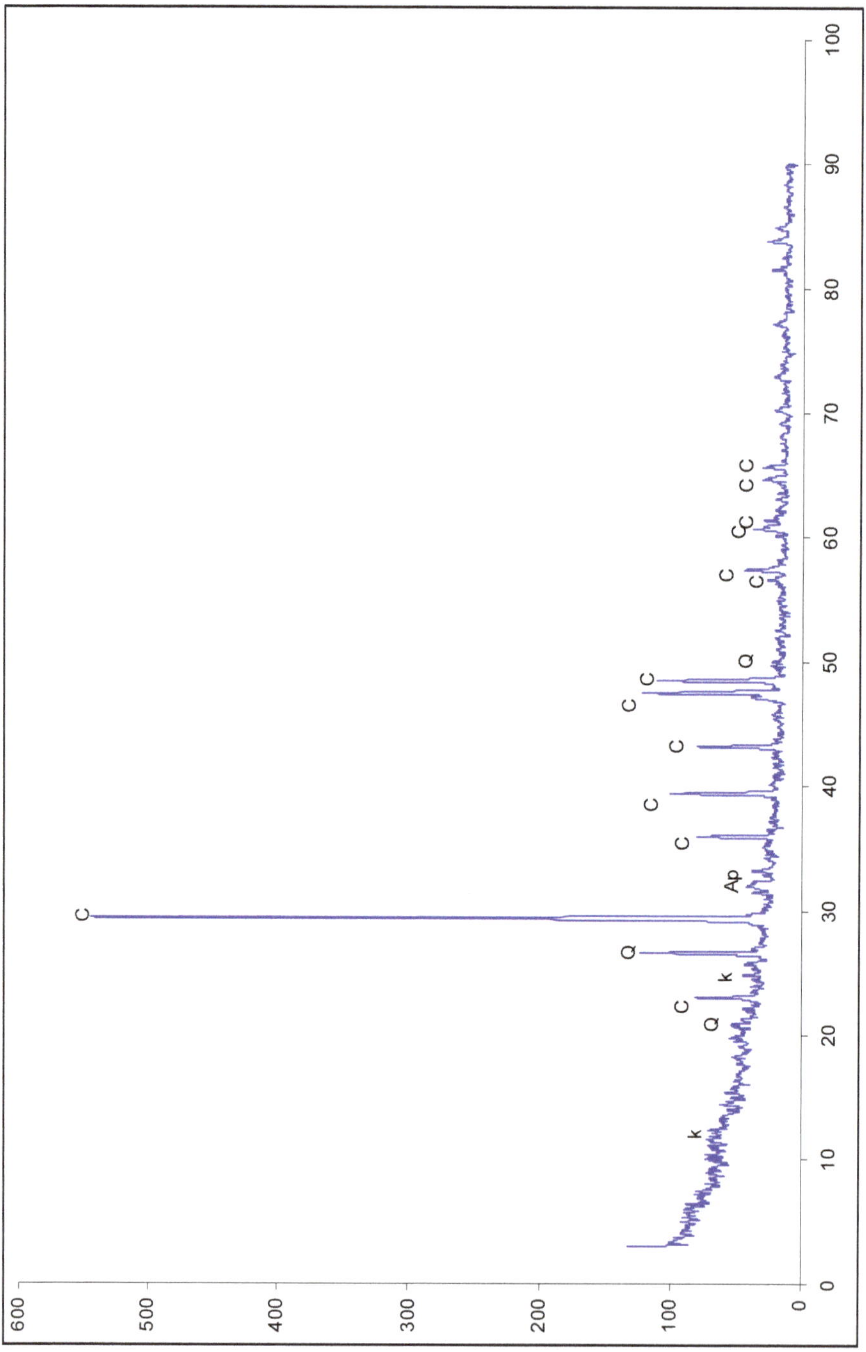

Figure 5.2: XRD Pattern of Untreated Oil Shale Sample.
(C: Calcite, Q: Quartz, K: Kaoline, Ap: apatite).

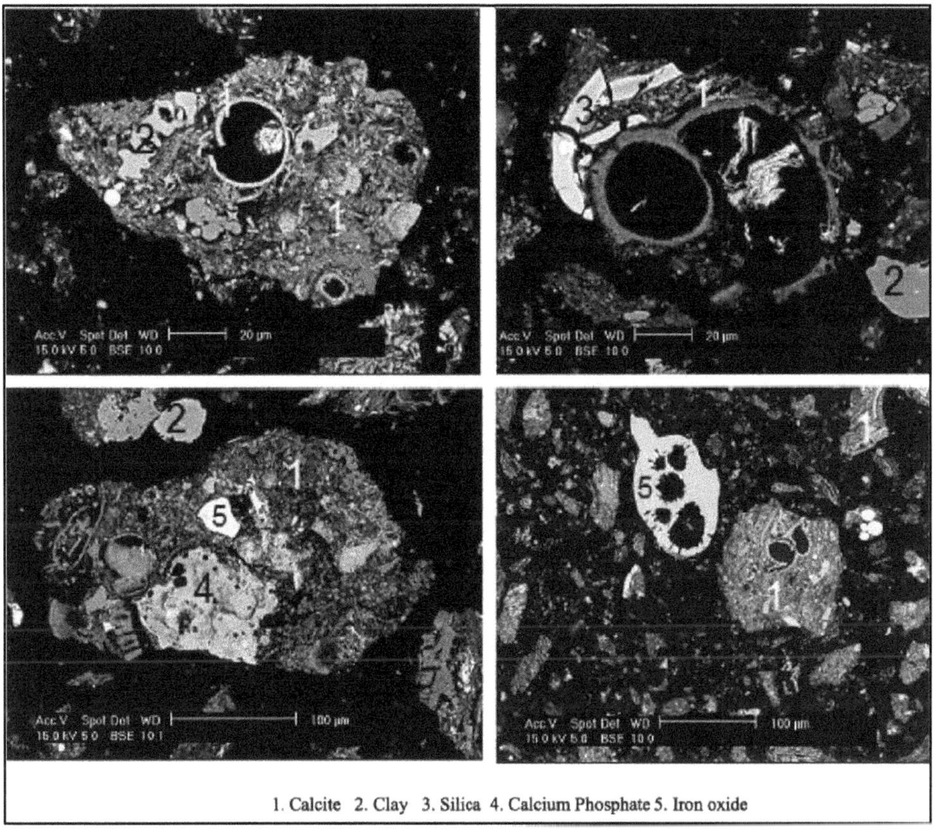

Figure 5.3: SEM for Oil Shale Showing different Phases of Minerals.

However, it was clear that there were some pure particles of calcite in the results of Qemscan. This information could help in the physical separation of kerogen rich minerals from pure minerals like oatation.

Froth Floatation of Oil Shale

A full factorial experimental design was performed in order to study the effect of different parameters on floatation of oil shale. As shown in Table 5.3, the major variables that believed to affect the oil shale recovery. Kerosene was used as a collector and MIBC was used as frother. The effect of particle size and the acidity of the solution have been also studied. Results from the table showed that none of these parameters enhance the oil shale recovery since the ash contents in the froth and tailing were almost constant at 54 per cent which is similar to the ash content in the original oil shale samples.

A further investigation on the type of the collector was performed to test the recovery of the organic materials. As shown in Table, changing the type of the collector did not change the recovery of oil shale in a significant magnitude. Slight improvement in the oil shale recovery was detected when using fuel oil #2 as a collector. The

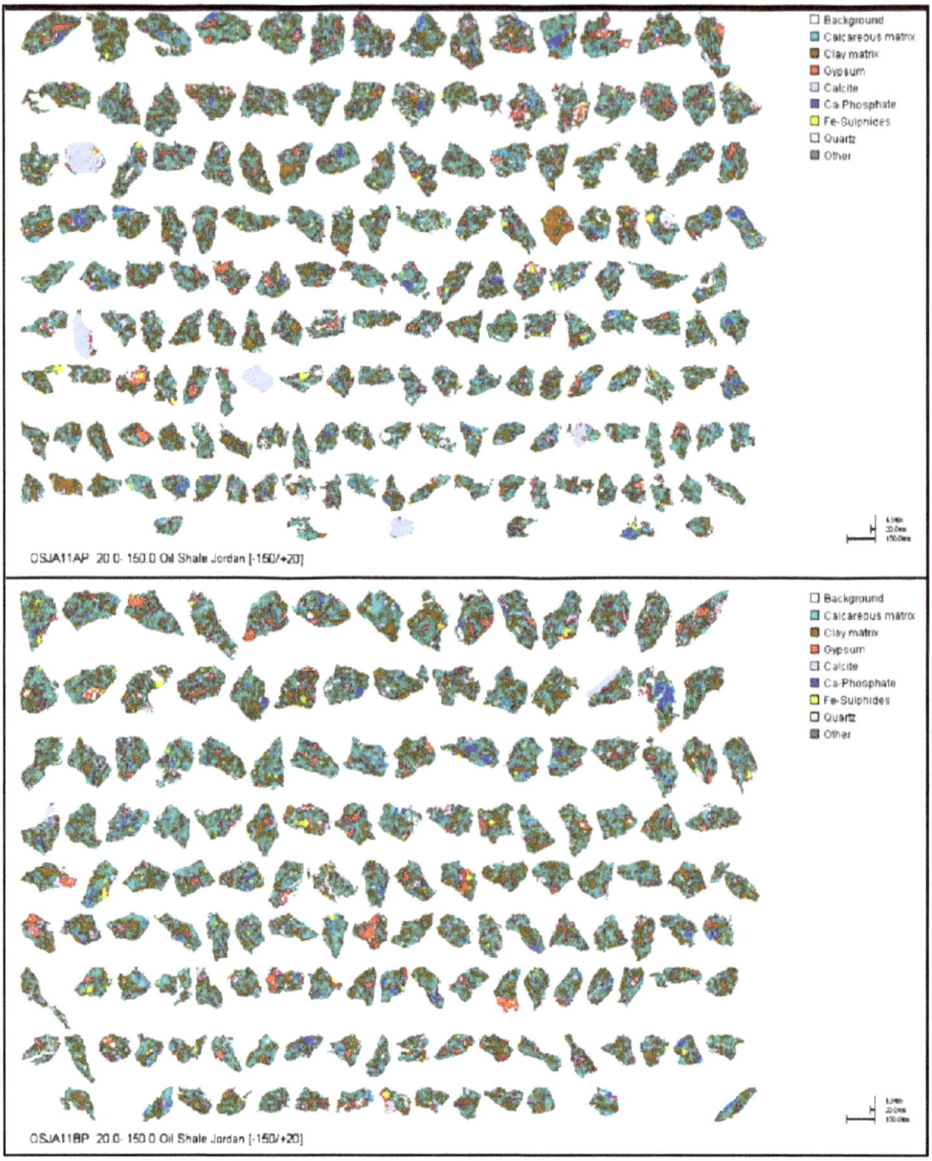

Figure 5.4: Qemscan Analyses for Two Samples of Oil Shale.

maximum enhancement was approximately around 3 per cent achieved in two stages froth floatation as shown in Table 5.5. It is considered a low recovery rate that might not justify investment in floatation units.

Although the recovery of the oil shale using fuel oil as a collector is relatively low, fuel oil is believed to attract more of the kerogen rich particles to the froth layer.

It was noted that around 73 per cent of original oil shale has moved to the froth side, while around 27 per cent moved into the tailing part. This can be explained by that during crushing of the oil shale samples, kerogen have wetted most of the oil shale and therefore has given most of the oil shale a hydrophobicity property. Since that hydrophobic material moves into the froth layer, it is expected that most part of the oil shale will move to that froth layer. This also can explain the low recovery of oil rich particles in the froth.

Table 5.3: Full Factorial Experimental Design of Floatation of Pure Oil Shale

Experiment	*Pattern of Variables*	*Mean Particle Size (μm)*	*Collector/ Frother*	*pH*	*Mass Per cent of Froth*	*Ash Percentage of Froth*
1	– - -	375	∝	4	73.8	54.6
2	– -+	375	∝	7	72.2	54.3
3	–+-	375	0	4	74.1	54.4
4	–++	375	0	7	73.3	53.89
5	+- -	55	∝	4	74.5	53.95
6	+-+	55	∝	7	73.8	53.7
7	++-	55	0	4	73.6	54.1
8	+++	55	0	7	74.6	53.87

Table 5.4: Effect of different Collectors on the Floatability of Oil Shale

Experiment	*Mean Particle Size (μm)*	*Collector*	*Frother*	*pH*	*Mass Per cent of Froth*	*Ash Percentage of Froth*
9	55	DMSO	MIBC	7	73.9	54.1
10	55	Sulphonated Naphthalene	MIBC	7	73.7	54.0
11	55	Sodium Lignosulfonate	MIBC	7	73.0	54.1
12	55	Calcium Lignosulfonate	MIBC	7	71.9	54.0
13	55	Diethyl Ketone	MIBC	7	73.9	54.6
14	55	Fuel oil #2	MIBC	7	72.9	52.8

Table 5.5: The Effect of Multiple Stages on the Floatability of Oil Shale

Experiment	*Mean Particle Size (μm)*	*Collector*	*Frother*	*pH*	*Number of Floatation Stages*	*Mass Per cent of Froth*	*Ash Percentage of Froth*
14	55	Fuel oil #2	MIBC	7	1	72.9	52.8
15	55	Fuel oil #2	MIBC	7	2	98.0	52.3
16	55	Fuel oil #2	MIBC	7	3	100	52.3

Demineralization of Oil Shale

The average oil yield of the original untreated oil shale from Ellajun deposit, as determined by Fischer Assay, was approximately 13 per cent which is consistent with earlier studies (Al-Harahsheh, 2005). The ratio of oil to gas is approximately 3.5.

Effect of HCl Digestion

Figure shows that the digestion with HCl produces a higher average total yield (30 per cent) than that obtained from untreated oil shale. The improvement in the total yield can be explained by the fact that hydrochloric acid can react readily and quickly with the calcium carbonate that exists in abundance in Ellajun oil shale according to the following reaction

$$CaCO_3 + 2HCl \Longrightarrow CaCl_2 + CO_2 + H_2O \tag{1}$$

This reaction results in a soluble calcium chloride which can be removed upon washing. This reaction will also dissolve the shell of calcium carbonate leaving the organic matter as it is mentioned earlier to be encapsulated in a shell of calcium carbonate.

The oil to gas ratio for treated oil shale with HCl, as presented inFigure 5.4, have shown a decreased oil to gas ratio. This can be explained by the fact that HCl can only digest the calcite leaving the clay, which has a great affinity to organic material. This hypothesis is also examined using the analysis of the XRD where as the concentration of HCl increases the intensity of the $CaCO_3$ peak in the digested sample decreases leaving the peaks of quartz and clay which are not soluble in the HCl solution.

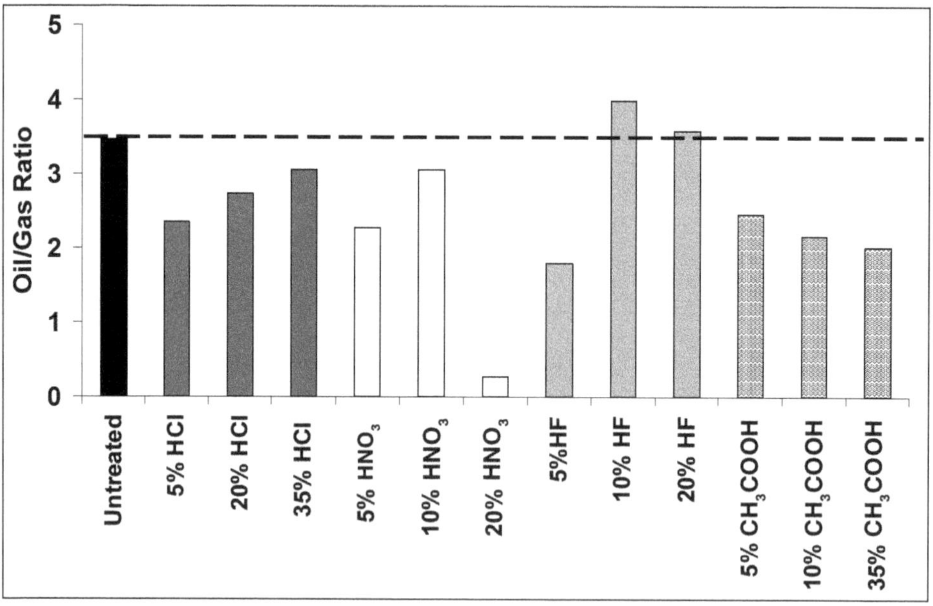

Figure 5.4: Oil/Gas Ratio Obtained by Fischer Analysis from Oil Shale Residue after Treatment with Various Acids.

Effect of HNO_3 Digestion

The average total yield obtained from the digestion of oil shale and nitric acid was approximately 28 per cent. Similarly the nitric acid reacts quickly with the calcium carbonate to form the soluble calcium nitrate according to the following reaction

$$CaCO_3 + 2HNO_3 \Longrightarrow Ca(NO_3)_2 + CO_2 + H_2O \quad (2)$$

Results of the XRD analysis for the HNO_3 digested oil shale have shown a decrease in the intensity of the calcium carbonate peak with increasing the concentrations of the HNO_3. Increased concentration of the nitric acid has produced a drastic decrease in the oil to gas ratio as shown in Figure 5.5. This could be explained by possible reaction between the nitric acid and the kerogen resulting in a lower molecular weight compounds that produce high amounts of gases during the Fischer Assay test.

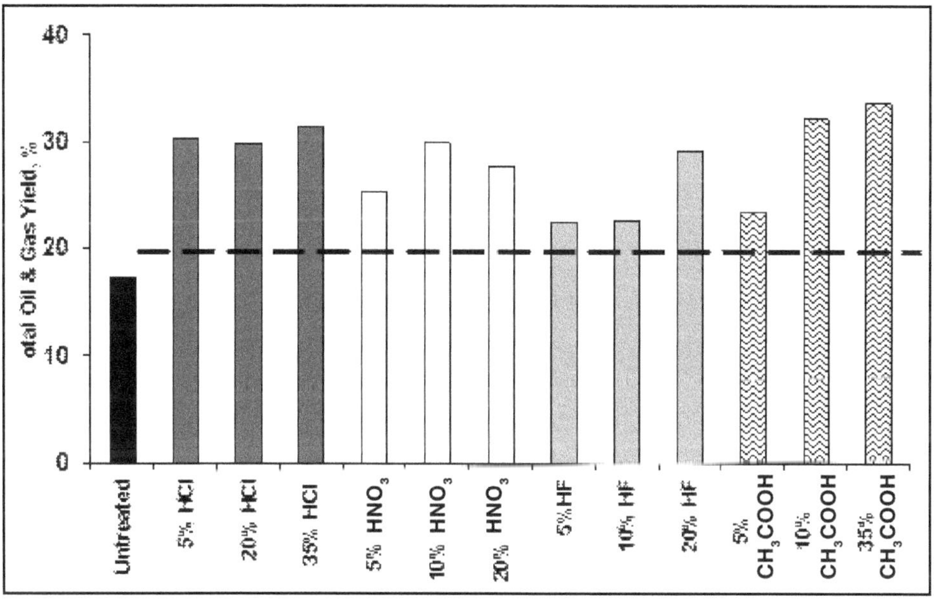

Figure 5.5: Total Yield Obtained by Fischer Assay for the Original Oil Shale and the Treated Samples.

Effect of HF Digestion

The treatment with hydrofluoric acid has resulted in an average total yield of 24.7 per cent. Hydrofluoric acid (HF) reacts also with the calcium carbonate producing insoluble calcium fluoride(CaF_2) according to the following reaction

$$CaCO_3 + 2HF \Longrightarrow CaF_2 + CO_2 + H_2O \quad (3)$$

At high concentrations, the hydroflouric acid appears to react with the kerogen forming a high level of low molecular weight of compounds. However, at low concentration of this acid, it is believed to be consumed by reacting with the calcite, therefore pH is increased and CaF_2 precipitate at this high pH. Once concentration of HF is increases, it seems that CaF_2 was not precipitated.

Effect of CH_3COOH Digestion

Results have shown that the digestion of the oil shale with acetic acid produces the highest total yield (33 per cent). Demineralization by acetic acid can be summarized by the following reaction with calcite

$$CaCO_3 + 2CH_3COOH \Longrightarrow Ca(CH_3COO)_2 + CO_2 + H_2O \quad (4)$$

It seems that when using high concentration of the acetic acid, partial transformation of kaolinite to illite occurred. Evidence about this transformation is the low water content in the oil obtained by Fischer Assay compared to that obtained when digesting with other acids. Illite would not produce any amount of water at the temperatures of pyrolysis by the fischer assay method (520°C) since the decomposition of illite occurs at 596°C (Jiang *et al.*, 2008). This transformation was evident in the XRD (Figure 5.6) for the oil shale treated with 35 per cent acetic acid. The transformation of kaolinite to illite was not evident in all other acids since kaolinite decomposes at a lower temperature (430°C) and certain amounts of water were condensed by the fischer assay method for all other acids.

Since the use of acetic acid has produced the highest total yield, the recovery of the acetic acid should be performed. The following reaction can show one method for the recovery of this acid

$$Ca(CH_3COO)_2 + H_2SO_4 + 2H_2O \Longrightarrow 2CH_3COOH + CaSO_4.2H_2O \quad (5)$$

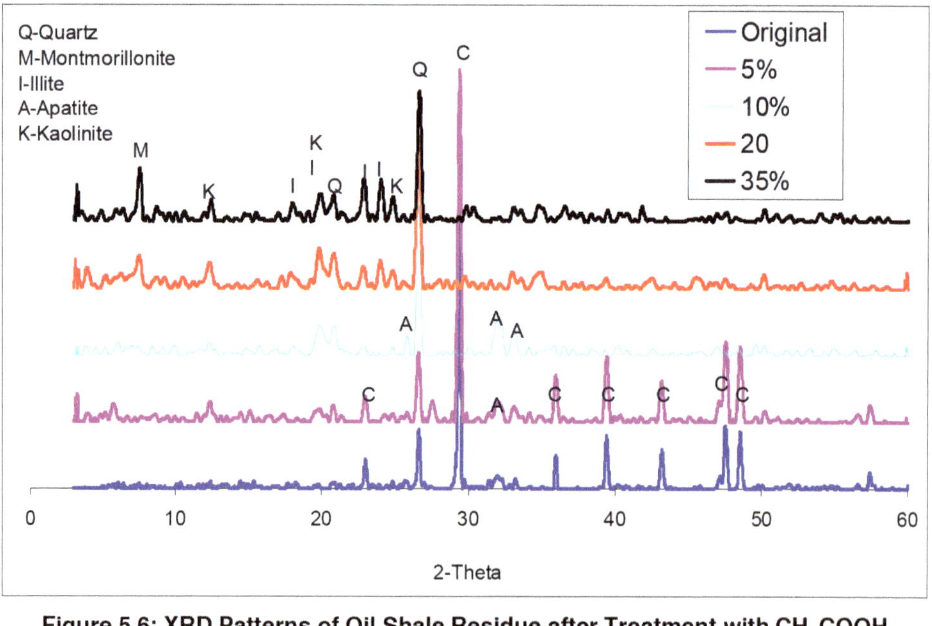

Figure 5.6: XRD Patterns of Oil Shale Residue after Treatment with CH_3COOH Solutions of various Concentrations.

Conclusions

During combustion, fragmentation of oil shale particles is minimal. The particle size distributions for both oil shale and oil shale ash are almost identical.

Different analytical techniques were used in this study to understand the transformation of minerals during combustion of oil shale. The dominant phase found in the oil shale was the calcite. Different proportions of clays, quartz, and phosphate were also present. There was no consistent distribution of minerals in oil shale particles. However, some pure calcite particles were also found in the sample of oil shale.

Sulfur was found to be mainly in the organic part of the oil shale. The organic part was found to be encapsulated by a shell of mineral, mainly calcite. Some inorganic sulfur was detected only by the Qemscan. During combustion, it is expected that the oxidation product of organic sulfur react with the calcite, which acts as a sulfur sorbent media, to produce the anhydrite.

Floatability of Jordanian oil shale has been studied and the effect of different parameters has been investigated. These include type of collector, pH level, particle size, and concentration of frother and collector. In addition, the separation of phosphate and calcite minerals has been investigated.

The floatation of oil shale has shown a poor recovery of oil rich components with a maximum of 3 per cent enhancement when using fuel oil #2 as a collector. This poor recovery is believed to be due to fact that kerogen is encapsulated in layers of minerals and that kerogen, during crushing and milling wets most of the oil shale particles making it difficult to adjust the surface properties.

Demineralization by treating Jordanian oil shale has demonstrated a technical viability and feasibility by increasing the oil yield by approximately two folds (17 to 33 per cent).

The demineralization of oil shale with hydrochloric acid has shown fairly constant total yield over a wide range of concentration by forming calcium chloride which is considered to have a high level of solubility in water.

The treatment with nitric acid at high concentration seemed to have reacted with koregen forming higher amounts of low molecular weight compounds. This has resulted in a dramatic decrease in the oil to gas ratio.

Acetic acid was the most efficient acid in the demineralization process as a result of reacting with most of the calcite forming calcium acetate. In addition, the treatment of oil shale with higher concentration of acetic acid seemed to have converted the kaolinite into illite which produced less amounts of water from the Fischer Assay tests.

References

1. Al-Harahsheh, A., A. Y. Al-Otoom, and R. A. Shawabkeh, 2005. Sulfur distribution in the oil fractions obtained by thermal cracking of Jordanian El-Lajjun oil Shale, Energy. 30, pp. 2784.

2. Dabbas, M. A, 1997. Oil shale: hopes and ambitions (in Arabic). in Second Jordanian Conference for Mechanical Engineering. Amman-Jordan: JAE.
3. Hamarneh, Y., J. Alali, and S. Sawaqed, 2006. Oil shale resources development in Jordan. Natural Resource Authority: Amman. p. 98.
4. Intellection Co.-Australia Qemscan. See also: http://www.intellection.com.au/scientic_papers.html (accessed 12 August 2004)
5. Jiang, T., G. Li, G. Qiu, X. Fan, and Z. Huang, 2008. Thermal activation and alkali dissolution of silicon from illite, *Applied Clay Science*. 40, pp. 81.
6. Khraisha, Y. H., 2000. Retorting of Oil Shale Followed by Solvent Extraction of Spent Shale: Experiment and Kinetic Analysis, Energy Sources. 22, pp. 347-355.
7. Vandenbroucke, M. and C. Largeau,2007. Kerogen origin, evolution and structure, Organic Geochemistry. 38, pp. 719-833.

Chapter 6

Overview of Mineral Processing and Beneficiation in Kenya

T.W. Githinji, C.M. Nyamai and A.W. Kutukhulu

Department of Geology,
University of Nairobi, P.O. Box 30197-00100 Nairobi
E-mail: tgithinji@uonbi.ac.ke

ABSTRACT

Kenya is endowed with enormous mineral resources both metallic and non-metallic minerals which are known and mapped. Mining activities are currently dominated by production of non-metallic minerals encompassing minerals such as soda ash, fluorspar, kaolin and some gemstones. Currently, mining accounts for a very small part of Kenya's annual GDP. It is therefore important for the country to put more effort to foster the establishment of a thriving mining sector.

Mining has continued in the well-known mineral deposits such as common salt, from the saltworks along the coastal Kenya and Lake Magadi, fluorspar in the Kerio Valley, and soda ash from Lake Magadi. One of the best-exploited industrial minerals, soda ash, is produced from the mineral trona that occurs at Lake Magadi, which is situated within the Great Rift Valley, Southern Kenya.

There are several challenges facing the mineral exploitation in Kenya. These include, lack of enough knowledge on the geology and geographical distribution of the minerals. This is mainly because not all parts of the country have been properly prospected. Secondly, the Mining Act (1940) is still outdated. Another reason is that there is a conservative emphasis on economic development focus on agriculture, tourism and wildlife conservation. Lastly, there is lack of enough technological knowledge on mineral processing and beneficiation.

The vast endowment of minerals in Kenya shows that the mining industry has a great potential to contribute more than 15 per cent of the national GDP. This projection can only be achieved if the miners, technologists and researchers worked on how to overcome the challenges and find better ways of enhancing mineral processing and beneficiation techniques. It is vital for mining industry to take up the issue of mineral processing seriously and urgently, so as to extend the knowledge to other minerals that could be of great value to the economy.

Keywords: *Overview, Geology, Mineral processing and beneficiation, Kenya.*

Introduction

Kenya is endowed with a lot of minerals such as iron, manganese, gold, soda ash (Trona), salt, fluorspar, ilmentile, rutile, zircon, diatomite, gypsum, vermiculite, manganese, laterite, pyrites, magnesite, limestone, lime, kaolin, silica and copper etc which are known and mapped. The mining industry has shown rapid growth in the recent years throughout the country. Moreover, the government has recognized the importance of the mineral sector in achieving of its vision 2030 for economic and industrial development. The government has drafted a mineral and mining policy that aims at creating an overarching framework for managing mineral resources in Kenya. The draft policy aims at facilitating the establishment of a vibrant mining sector that can attract substantial investment for research, exploration and exploitation of mineral resources in the country.

The government is also training more expertise in the mineral industry and more funds have been channeled in the exploration of minerals. It is also trying to create conducive environment for the international investors in the mining sector. It is also encouraging the collaboration between the researchers and academicians with the miners.

Globally the demands for minerals as a raw material for various industries have increased at a very high rate. This has consequently led to the increase in exploration and exploitation activities throughout the country. Though the Kenyan government is working hard toward the development of the mining industry, it still faces the following challenges:

1. Lack of adequate finances to carry out large scale mineral exploration programmes,
2. Few Kenyan expertise in the mining industries,
3. Large tracks of the country are still under explored
4. Favorable mining act policy has not yet been implemented.
5. There is little mineral processing and beneficiation of the exploited minerals.
6. There is limited collaboration between research institutions and the mining prospectors.

For decades the country has been exploiting fluorspar in Kerio valley and Trona in Lake Magadi, in the Rift Valley. These two commodities have been the Kenyan largest income earners in the mineral industry. There are other minerals that have been mined in small scale such as gemstones, gold, iron etc. The unfortunate thing is that, the bulk of the exploited minerals are exported without processing and beneficiation. Consequently, the mining sector has a stagnant small contribution to the country's annual GDP stands at less than 1 per cent (Table 6.1).

Table 6.1: Contribution of the Mining Sector to Kenya's Gross Domestic Product (GDP); Data from Kenya National Bureau of Statistics (2012)

National Accounts

Percentage Contributions to GDP by Activity (Current Prices)

Industry	*2008*	*2009*	*2010*	*2011*
Agriculture and forestry	22.3	23.5	21.4	24.0
Fishing	0.4	0.4	0.6	0.5
Mining and quarrying	0.7	0.5	0.7	0.7
Manufacturing	10.8	9.9	9.9	9.4
Electricity and water supply	2.1	1.9	1.4	0.9
Construction	3.8	4.1	4.3	4.1
Wholesale and retail trade, repairs	10.2	9.8	10.2	10.6
Hotels and restaurants	1.1	1.7	1.7	1.7
Transport and communication	10.3	9.9	10.0	9.7
Financial intermediation	4.6	5.4	5.6	6.4
Real estate, renting and business services	5.1	4.9	4.8	4.5
Public administration and defence	5.0	5.0	5.6	5.0
Education	6.3	6.0	6.2	5.8
Health and social work	2.4	2.5	2.6	2.5
Other community, social and personal services	3.4	3.4	3.3	3.2
Private households with employed persons	0.4	0.4	0.4	0.4
Less: financial services indirectly measured	–0.9	–1.1	– 0.8	–1.1
Taxes less subsides on products	11.8	11.7	12.2	11.7
GDP at market price	**100**	**100**	**100**	**100**

The current contribution to the GDP could be increased, even up to 15 per cent. This can only happen if the country addresses the challenges facing the mining sector in Kenya. This paper discusses the current mining and mineral processing in Kenya and intends to address on how to improve the mining sector.

Legal Framework Covering Mining in Kenya

Mineral exploration and exploitation are currently carried under the auspices of the Mining Act, *Cap.306* of the Laws of Kenya, and is administered by the Department of Mines and Geology, Ministry of Environment and Natural Resources. The Department has the responsibility of undertaking geological surveys, geoscientific research, coordination and regulation of the activities of the mining sector.

In Section 2 of the Mining Act, it defines minerals as precious metals, precious stones or non-precious minerals, but does not include mineral oil, clay, murram, limestone, sandstone, mineral water, brine, dolomite, kaolin, gravels, dimension stones, Kisii stone (pyrophyllite), sodium and potassium compounds except those forming part of Lake Magadi saline deposit and ornamental stones in economically workable quantities.

All un-extracted minerals under or upon any land, as per the Mining Act *Cap 306* of Kenya, are vested in the Government, subject to any rights which under the Act, can be granted to any other person. Currently the Government policy in mineral resources development is to ensure that the private sector takes the leading role in mineral development while it assumes a promotional, encouragement and regulatory role by providing basic geological data and necessary fiscal incentives. The Government of Kenya undertakes reviews of policies, mineral legislation, promotion measures and publicizes the mineral investments in Kenya through the Mines and Geology Department in the Ministry of Environment and Natural Resources (Government of Kenya, 1940).

The commissioner of mines and geology under the mining act *Cap 306* is mandated to grant licenses as; (1) Mining Location; (2) Exclusive Prospecting License (EPL); (3) Special License;(4) Mining Lease; (5) Special Mining Lease. Mining locations are mostly granted to small-scale miners. The conditions attached to their grant are less onerous than in the case of other titles. A location consists of a block of not more than ten claims. A claim is an area of: (a) 20,000m^2 (or 200 metres x 100 metres) in the case of precious metals (gold, silver etc) and precious stones (ruby, emerald, opal etc). (b) 50,000 m^2 (or 250 metres x 200m) for all other minerals (copper, lead, graphite, baryte etc). A Mining Lease or a Special Mining Lease is granted for the exploitation of a mineral deposit, which has been discovered following successful prospecting and exploration under the first three types of licenses.

An Exclusive Prospecting License or Special License is granted over an area selected by the applicant who has a valid Prospecting Right, has deposited a sum of 2000 Kenya shillings' with the Provincial Commissioner of the Province in the area over which the application made is situated, has written consents from relevant local authorities, has consents from land owners, etc. Mineral exploration, exploitation and trade are presently carried out under the auspices of the Mining Act *Cap 306*, of the Laws of Kenya, the Trading in Unwrought Precious Metals Act *Cap 309*, of the Laws of Kenya and the Diamond Industry Protection Act *Cap 310* of the Laws of Kenya (Government of Kenya, 1940).

In order to facilitate the establishment of a vibrant mining and mineral sector in Kenya, the mining policy have been reviewed, the aim of this policy is to advocate for optimal utilization of mineral resources and sound environmental management in order to ensure sustainable development of the sector. The reviewed draft mineral law is to provide for lesser discretionary powers to the licensing authorities and hence provide for greater security of tenure. It also aims to provide for greater environmental protection, as embodied in the 1999 Environment Management and Coordination Act, from the undesirable effects of mining activities.

Geology and Distribution of Minerals in Kenya

The geology of Kenya is characterized by Archean granite/greenstone terrain in western Kenya along Lake Victoria, the Neoproterozoic 'Pan-African' Mozambique Belt, which underlies the central part of the country and Mesozoic to Recent sediments underlying the eastern coastal areas. The Eastern Rift Valley crosses Kenya from north to south and the volcanics associated with rift formation largely obliterate the generally north-south striking Neoproterozoic Mozambique Belt. Rift Valley volcanogenic sediments and lacustrine and alluvial sediments cover large parts of the Eastern Rift.

The main mineral geological sources are;

(a) Archean Nyanzian Shield Area of Western Kenya

The Nyanzian system is mainly composed of lavas and pyroclastics with minor sediments and banded ironstones. The Kavirondian, which rests unconformably on the Nyanzian, consists of grits, sandstones, greywackes and conglomerates. Both the Nyanzian and Kavirondian systems (Figure 6.1) are isoclinally folded about axes that have an east-westerly trend. Kavirondian is only slightly younger than Nyanzian but folding in the two systems, has similar orientation. Numerous granitic bosses and batholiths have intruded the Nyanzian and Kavirondian. The Kavirondian intrusions were more but the pre-Kavirondian were also widespread and the two systems are discernible. In the northern part of Nyanza Province the main granites are post-Kavirondian and genetically connected with the gold deposits. In South Nyanza the principal intrusion has, however, been considered to be of post-Nyanzian-pre-Kavirondian age, and is apparently associated with gold mineralization in the 'Migori gold-belt. Metallic mineralization is common and has potential for Ferrous and ferrous metals.

(b) The Proterozoic Mozambique Belt

It is a structural unit within which a wide variety of meta-sedimentary and meta-igneous rocks are found showing a broad concordance of structural style and metamorphic history. It is most extensive in Kenya Central, North to South (Fig. 1) in which metamorphic minerals such as Kyanite, Corundum, Graphite, Wollastonite, Marble, kaolin dand a variety of gemstones are found together with minerals associated with basic and granitic rocks, such as mica and iron ore.

(c) The Sedimentary Rocks

These are also widespread and range from Palaeozoic to Recent. Palaeozoic and Mesozoic formations in Kenya are found near the coast and in the north-eastern part of Kenya. The earliest of these rocks are Permo-Carboniferous which are composed essentially of sandstones and shales that form the Duruma series. This is equivalent to the Karroo system in Southern Africa. The local formations are Taru; Maji-ya-Chumvi; Mariakani and the Mazeras. They extend for about 100 kilometres from Taru to Mazeras, west of Mombasa. The rocks dip very gently towards the ocean and are heavily faulted in places. Mesozoic rocks occur in two separate areas, in the north-east part of Kenya and along the Coast belt, Horkel *et al.* (1984). They are

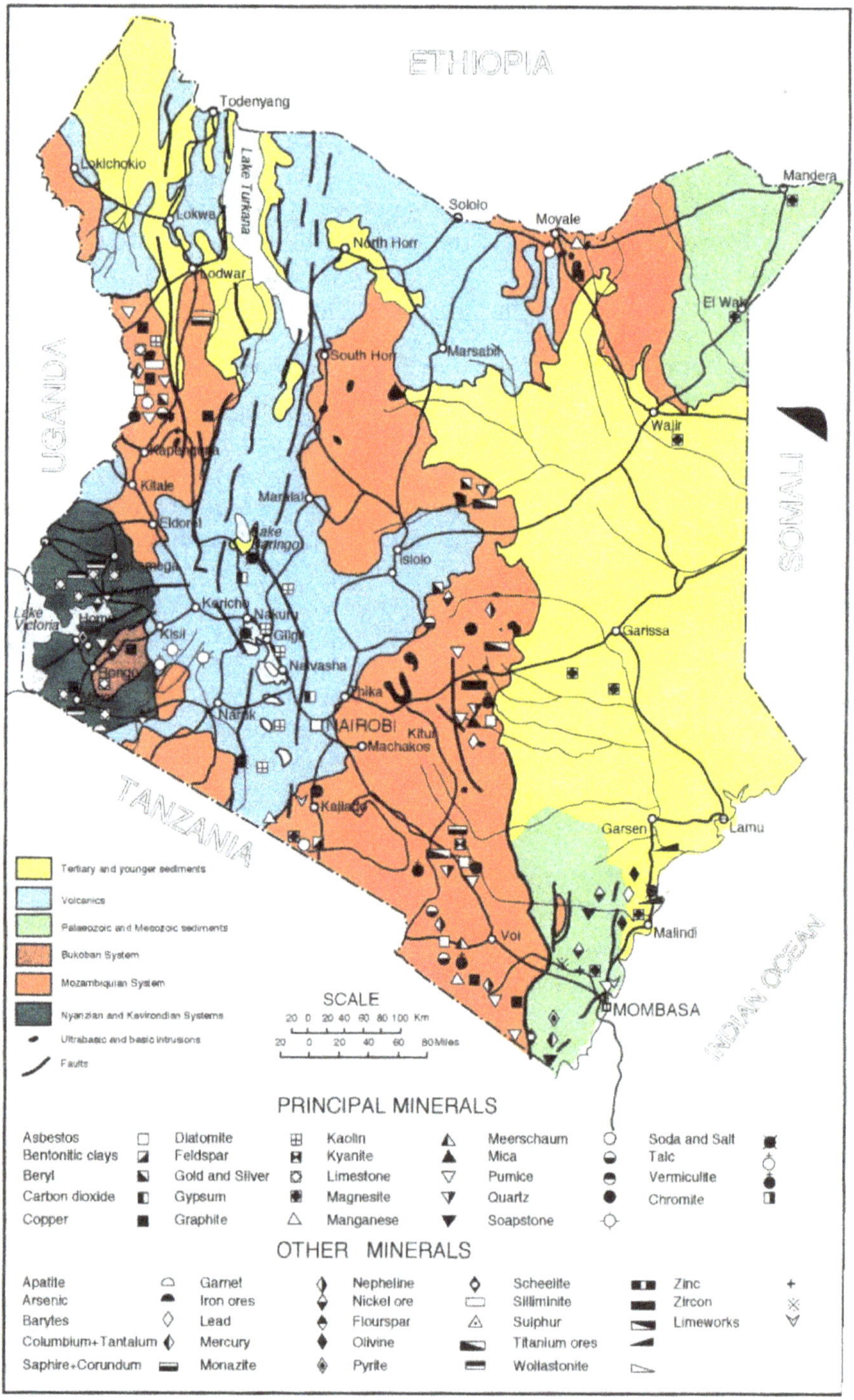

Figure 6.1: Map Showing Geology and Mineral Occurrence in Kenya (Modified from mines and geology map 1962).

possible sources and hosts of hydrocarbons, limestone, gypsum, clays, and manganese. Base metal mineralization, lead-zinc is known to occur in the sedimentary basin along the Coastal belt. Heavy mineral sand also occurs along the Coastal beach sand according to Walsh (1970).

(d) The Volcanic rocks

Associated with the rift system host and yield a variety of minerals and construction materials. These volcanic-Sedimentary accumulations have deposits of Clays, Trona, Diatomite, Natural Carbon dioxide, Kunkar and Gypsum. The geothermal fields are found in this area (Davies and Mathu, 1996). Geothermal electricity generation, which has a potential of 2,200MW, is currently undergoing exploitation at the Kenya Rift Valley province.

(e) The Mrima Ore

The Mrima Ore the Carbonatite which has a potential for niobium and other rare earth elements (REE) is found in the coast basin south of Mombasa. This ore has been found to contain large deposits of rutile, ilmenite and zircon that also contain titanium. According to Straaten (2002), Phosphates at Mrima Hill occur mainly in the form of monazite (a Cerium-Lanthanum phosphate mineral) and the complex barium phosphate mineral gorceixite. Only very minor amounts of apatite have been reported from Mrima Hill and no occurrences of secondary calcium-phosphates, such as francolite are reported.

Current Mining and Mineral Processing Activities in Kenya

The mining of industrial minerals dominates the Kenyan mining sector in terms of income. Among the major contributors to the mining industry in Kenya include the following minerals:

Trona (Soda Ash)

Soda ash from Lake Magadi in the Great Rift Valley in southwestern Kenya, accounts for 74 per cent of the total value of mineral exports, with an estimated production of 450,647 Mt in 2010 compared to 419,679 Mt in 2011 (Table 6.2). Tata Chemicals Limited is the current owner of the company and is the sole producer of soda ash in Kenya.

Soda ash or anhydrous sodium carbonate (Na_2CO_3), the main product of the Company, is formed during the calcining or heating of the natural crystalline carbonates of soda. The principal carbonate passing into the calciners at Magadi is sodium sesquicarbonate and this loses water of crystallization and breaks up into soda ash in accordance with the following reaction:

$$2(Na_2CO_3.NaHCO_3.2H_2O) \rightarrow 3Na_2CO_3+5H_2O+CO_2$$

The end products of the reaction include steam and carbon dioxide gas which pass out of the calciners with the hot flue gases and escape from chimneys above the ash plant. The company mines trona from Lake Magadi, which has one of the purest surface deposits of trona. The trona is processed to soda ash, at a facility near the mining operations, and the soda ash is transported by rail to Mombasa for onward

Table 6.2: Mineral Production in Kenya for the Period 2006–2011

Mineral	Unit	2006 Quantity	2007 Quantity	2008 Quantity	2009 Quantity	2010 Quantity	2011 Quantity
Soda Ash	MT	316,077	307,289	389,100	365,358	419,679	450,647
Gold	Gms	432,371	3,023,325	339,923	1,055,330	2,355,176	1,635,597.03
Fluorspar	MT	132,030	85,115	130,100	5,500	40,750	95,051.01
Diatomite	MT	185	201	72	231	224	712.50
Crushed refined soda	MT	5,085	54,695	38,280	16,331	10,710	2,857
Salt	MT	837	2,109	1,106	592	449	3,327
Carbon dioxide	MT	5,048	5,645	4,900	5,460	1,127	–
Gemstones	Gms	50,439,006	119,286,344	20,933,983	39,408,772	167,625,352	310,096,571.43
Iron ore	MT						114,800
Other minerals		95,012	90,020,643	67,486	21,969	27,827	

NB. Data from Mines and Geological Department, Kenya.

shipping (Baker 1957). The bulk of the commodity is exported but some is used locally for the manufacture of various industrial materials.

Fluorspar

Fluorspar is the second largest mineral income earner in the country; It is mined by Kenya Flouspar Company Ltd. The mineral deposit is located East of Eldoret town in the Kenya Rift valley. Fluorspar is primarily mined for export. The mined ore is hauled a short distance of between 2 km and 7 km by contracted trucks to the crushing plant. The ore is first crushed in a tertiary stage crushing plant from a feed of 350 mm to less than 16 mm in size. It is then conveyed to the grinding circuit where the material is added to a rod mill along with water to form slurry. The rod mill instantly grinds the ore after which it is sized using a hydrocyclone. Any oversized material is passed to the ball mill for a further grinding to the desired size of 80 per cent. The ore is now ready for flotation, Kenya Fluorspar Company (2012).

Flotation is the process that concentrates the ore. This is done by agitating the ore slurry in cells with air bubbles. By adding a combination of fatty acid reagents, the fluorspar in the ore attaches itself to the air bubble to float to the top of the cell. This product is skimmed off leaving the waste in the bottom of the cell.

The process is conducted in a series of "rougher" and "cleaner" cells that successively concentrates the ore from 40 per cent CaF_2 in the feed material to a minimum of 97.0 per cent CaF_2 in the final concentrate. The water in the final product is then removed in a thickener and a rotary drum filter. This produces a filter cake concentrate containing approximately 11.0 per cent moisture. The samples are analyzed in the company's assay and research laboratory. The waste product is pumped to the tailings dam and settled water is recycled to the plant for reuse Kenya Fluorspars Company (2012).

From fluorspar the company also produces fluorite, which is a key component in the manufacture of some industrial commodities, including steel, hydrofluoric acid and opalescent glass. There was an estimated production of 95,051 Mt of fluorspar produced in 2011 compared to 40,750 Mt in 2010 (see Table 6.2).

Diatomite

Diatomite occurs at many points within the Rift Valley System and is mined only at Gilgil for export and local use, with reported production of 712 mt in 2011 compared to 224mt in 2010. Exploitation of diatomite has continued at Soysambu and Kariandusi of the Nakuru district. The demand for diatomite has been significantly reduced due to the closure of some companies that were using diatomite in refining their products.

Vermiculite

Vermiculite is found in the Pre-Cambrian rocks of the Mozambique Belt, and in the north central part of the Rift Valley province. One of the vermiculite deposits is located at Kinyiki Hill, half way between Nairobi and Mombasa. This vermiculite mineralization, associated with an ultramafic dunite body, was exploited in the past. However, the volume was small and mining ceased. A vermiculite deposit that has been mined intermittently is the Wasin Mine near Lodosoit, approximately 30 km west of the main Nanyuki-Isiolo-Marsabit road in northern Kenya. The vermiculite mineralization is associated with an altered sheared mafic rock (Walsh 1970; Mason and Theuri 1980). Between 1972 and 1998 some 72,000 tonnes of vermiculite were produced from the Wasin mine. Peak production was reached in 1987 when 9,220 tonnes were produced at Wasin. Most of the vermiculite products are used in the local building industry. Only minor use of vermiculite is made in the horticultural industry of Kenya.

Gypsum

Gypsum is found in the Tana River along the coastal belt, around Konza and in Turkana to the Northwest and is mined primarily for sale to local cement plants. Several gypsum occurrences are reported from the eastern part of the country and two occurrences from central Kenya (Figure 6.2). Impure gypsum is reported from Garissa consisting of pinkish gypsum in clays. Indications of large volumes of gypsum in soft earthy gypsum/clay aggregates (approximately 2,000 million short tons) are located near El Wak at the Somali border (Walsh 1970). Good quality gypsum occurs in the Kajiado area in central Kenya and is currently being mined and used for the cement industry. Some of the gypsum used as a redundant for the Bamburi cement industry near Mombasa is mined at Roka, 37 km southwest of Malindi (Theuri 1994). Gypsum used in cement industry is crushed to sizable quantities, while the one used in the manufacture of plaster of paris is calcined.

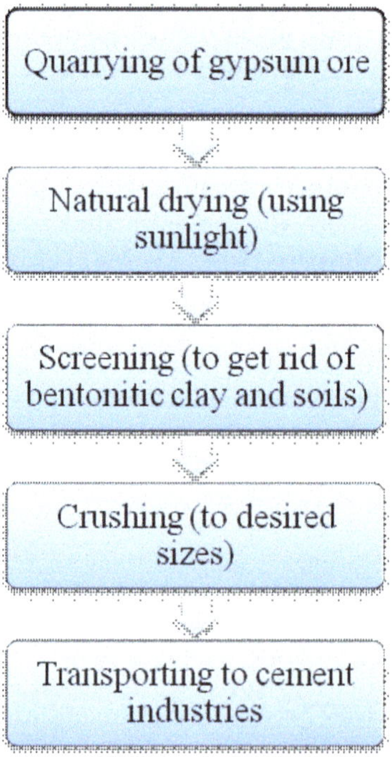

Figure 6.2: Schematic Diagram Showing the Mining and Processing of Gypsum.

Gold

Gold production in Kenya is mainly artisanal, the small scale miner process and sell their gold locally to businessmen. Goldplat is currently acquiring the remaining 50 per cent of the Kilimapesa Gold project from International Gold Exploration AB. The project is located in the historically producing Migori Archaean Greenstone Belt in western Kenya, which includes the operating Kilimapesa gold mine and adjacent exploration assets comprising Licence 91B.Kilimapesa Gold commenced initial production in January 2009 and has announced an initial JORC-compliant resource of 1.65Mt at 2.44 g/t Au for 129,000 oz Au at a cut-off grade of 1 g/t Au for all categories.

Heavy Metals

Base Resources Limited is developing the world class Kwale Mineral Sands Project in Kenya. The minerals include titanium, ilemnite, zircon and rutile. Table 6.3 lists the Kwale Ore Reserves by classification and dune. The Ore Reserves are estimated using all available geological, relevant drill hole and assay data, including mineralogical sampling and test work on mineral recoveries and final product qualities. The Ore Reserve estimates are determined by the consideration of all of the modifying factors in accordance with the JORC Code 2004. According to this code, this may include but are not limited to product prices, mining costs, mining dilution and recovery, metallurgical recoveries, environmental considerations, access and approval. The mineral assemblage is reported as a percentage of insitu ore.

Table 6.3: Ore Reserve Statement for Kwale Mineral Sands Project in Kenya

Dune	*Classification*	*Tonnes (Mt)*	*THM (%)*	*Slime (%)*	*Oversize (%)*	*Ilmenite (%)*	*Rutile (%)*	*Zircon (%)*
Central	Proven	46.3	6.9	24.6	0.4	3.93	0.91	0.42
	Probable	29.2	4.5	24.5	1.0	2.45	0.61	0.26
	Proven and Probable	75.5	6.0	24.6	0.7	3.36	0.8	0.36
South	Proven	39.9	3.7	26.5	1.7	1.89	0.52	0.22
	Probable	25.2	3.4	29.2	4.8	1.42	0.4	0.17
	Proven and Probable	65.1	3.6	27.6	2.9	1.71	0.48	0.2
Total	Proven	86.2	5.4	25.5	1.0	2.99	0.73	0.33
	Probable	54.4	4.0	26.6	2.7	1.97	0.51	0.22
	Proven and Probable	140.6	4.9	25.9	1.7	2.59	0.65	0.29

A series of metallurgical test programmes, including pilot plant test-work and closed-circuit trials, were conducted between 1997 and 2005, culminating in the design of a wet concentrator plant ("WCP") and a mineral separation plant ("MSP") in 2006 as part of Vaaldiam's DFS.

Base Resources Limited reviewed the previous testwork and plant design and determined that potential existed to simplify the circuits by:

- Converting the desliming circuit from two-stage cyclone separation to a single-stage cyclone plus constant density tank separation circuit;
- Using more modern mineral separation equipment to simplify the ilmenite circuit;
- Removing wet high intensity magnetic separation ("WHIMS") from the ilmenite circuit, which also removed the need for ilmenite stockpiling and drying stages; and
- Using a series of high tension ("HT") electrostatic separators and rare earth roll magnetic ("RER") separation stages to simplify the rutile circuit.

Testwork at Allied Mineral Laboratories was performed on a sample of the heavy mineral concentrate ("HMC") produced from the 2004 pilot plant operation. This testwork confirmed the performance of simplified circuit design and the revised flowsheet was used as the basis for the redesign of the Mineral Separation Plant ("MSP") conducted during the EDFS, Base recourses limited (website).

Gemstones

Kenya boasts of a wide range of colored and ornamental stones which are mined in the country. These mainly include, in order of importance, ruby, tsavorite, sapphire, and various types of garnet, tourmaline, aquamarine and others. These gemstones are mined and the bulk is exported while some are cut and sold locally for jewellery.

Mineral processing and beneficiation of gemstone in Kenya is improving, there is more efforts exploration, and research on geographical and geological occurrences of colored stones in Kenya. Some of Kenya's gemstones and precious minerals include green garnet, ruby corundum, tsavorite, ruby, zoisite, sapphire, tourmaline and aquamarine.Ruby is currently being mined in Taita Taveta, Kwale, Kitui, West Pokot and Baringo districts; tsavorite in Taita Taveta Kwale districts; sapphire in the Kitui, Mwingi and Isiolo districts; tourmaline in Taita Taveta, Kajiado, Kwale and Kitui districts and aquamarine in Meru and Tharaka districts.

The gemstone and jewellery sub-sector in Kenya is liberalized with no trade restrictions. So as long as a miner or dealer is licensed, he or she can trade in an otherwise free market. Dealers can sell their products either as raw material or as finished products although the value addition process is highly encouraged.

Other Minerals

Apart from the main mining firms there are over 150 small scale operators and more than 20 small to medium scale exploration and mining companies operating in a total concession area of about 20,000 km^2 or slightly more than 3 per cent of Kenya's surface area.

Challenges Facing Mining, Mineral Processing and Beneficiation in Kenya

Insufficient knowledge on the geology and geographical distribution of minerals is a major problem in Kenya. Most of geological mapping was done by early researchers

Table 6.4: Other Minerals in Kenya

Mineral	*Occurrence*	*Utilization of the Product*
Apatite	Kitui	Manufacture of fertilizers
Diatomite	Gilgil	Filtering agent
Copper	Kitui/Nyanza	Manufacture of electrical appliances, Bulk ore exported
Galena	Vitengeni	Source of lead
Magnesite	Taita	Bulk ore exported
Bentonite	Athi river, Matoni hill	Emulsifying agent for asphaltic and resinous substances, ceramic
Kaolin	Karatina	Paper
Coal	Mui basin	Steel industry/energy source
Calcite	Coast province	Manufacture of Cement
Kyanite	Machakos	Furnace bricks

throughout Kenya, in 1950s and 1960s. During this time, there were no improved technology for mapping, and most of the exploration was carried out by traverses and collection of the fresh rock sample exposed on the surface. Therefore, the level of geological exploration acquired so far is inadequate for the evaluation of the country's mineral potential. The small scale miners in most parts of Kenya cannot access information found at the Mines and Geological Department library easily and are not in a position to interpret them. As a result they base their explorations on pure lack, which can sometimes cost a fortune.

In the recent past, the country has experienced a rise in the exploration and mineral development projects; this is due to the global upsurge in demand of minerals as raw materials for the industrial and economic development. Consequently, this has inhibited extensive exploration due to the increased competition between the mineral explorers, mining companies and land owners.

In the Mining act *Cap 306* of Kenya (1940), the mining legislation does not offer a conducive investment environment to potential and existing investors Kenya has not had a well defined land policy over the years. This, together with the existence of many land laws, some of which are incompatible with the various land uses in the country, has caused complexity in land management and administration. There are conflicts that emanate when a mineral right holder conducts all prospecting and exploration works and is henceforth denied access to mine or access auxiliary infrastructure by a landowner. The landowner may then start exploiting the minerals putting the miner at a great loss. This greatly discourages mining investments in Kenya due to lack of security of tenure for mining rights.

Section 7 (1) (m) of the current Mining Act (1940) and section 22(1) of the Mining and Minerals Bill 2009 stipulate that no prospecting and mining operations will be conducted in private lands, except with the consent of the owner thereof. It is physically impossible for a company undertaking large scale operations to seek for consent of potentially thousands of land owners for a reconnaissance or exploration license.

It is also very challenging to apply the best technology on mineral exploration, mining, mineral processing and beneficiation. This is mainly attributed to by the high cost of investments. There is also lack of adequate collaboration between the research institutions and the miners. This limits the transfer of knowledge on the new innovation and invention on mineral processing and beneficiation techniques.

Gold in Kenya in mainly exploited by that artisans who suffer challenges such as lack of capital, lack of cheap and rapid response consulting experts; lack of basic knowledge on identification of hosting minerals; inability to give reliable estimation of amount of mineral reserves; lack of knowledge of most appropriate technology for optimum yields; Davies and Osano (2005) lack of processing facilities; lack of training in the understanding of Earth processes; lack of reliable and well-paying markets; loss of mineralized land to big companies; and ignorance of mining regulations and laws.

Recommendation

The increase of the contribution of mining industry to the national GDP and enhancement of sustainable development is dependent on:

1. Enactment of the mining policy by the government: The government should speed up the implementation of the new mining policy which will encourage local and international investors.
2. Intensive mineral exploration activities: a revision-mapping programme should be pursued in order to improve geological knowledge of the country. A 1:50,000-scale mapping and aerial geological surveys are most appropriate for the acquisition of adequate information. The department of mines and geological survey in Kenya should also improve its cadastral maps and upload them on their website so that they are easily retrieved by interested parties. Geological information is critical to the mineral sector growth and expansion. Up to date information is fundamental in identifying exploration targets, to providing the government with the capacity to effectively promote and license national exploration and mining activities. It is important therefore that existing data be upgraded through digitization in order to create favorable conditions for evaluation. Easily retrievable information would facilitate immediate identification of potential areas and delineation of endowed areas. This will drive faster investment in exploration and mining.
3. Mineral processing and beneficiation: Kenyan should take up the responsibility of adding value to the mined minerals by processing them before export. Most of the wealth from the minerals is exported to the developed nations and thus Kenyans do not benefit from these resources. The government should also create conducive environment for the establishment of productive and sustainable mineral processing industry in Kenya.

4. The artisanal and small scale miners should receive the support in their mining operations. Proper consultation on occurrences and distribution of mineral should be availed to them. They should also be facilitated with proper processing facilities and market for their products.
5. There should be collaboration between the researcher, academicians and the miners in the field to enhance better mineral exploration strategies.
6. Development of reliable infrastructure that will support mining sectors. The government should construct roads to open up remotes areas where most of the mineral deposits occur.

Acknowledgement

I would like to express my special gratitude to the NAM S&T Center for the chance to present this paper, I am very grateful to the Ministry of Higher Education Science and Technology for offering me the opportunity and the facilitation, not forgetting the ministry of Environment and Natural recourses for the availing the important data necessary for the preparation of this paper. Special thanks to the Department of Geology, University of Nairobi for the facilities, time and resourceful insights from my fellow authors during the preparation of this paper.

References

1. Baker B.H (1957): Geology of Magadi area Government of Kenya report pp. 86.
2. Base resources limited projects on kwale-mineral-sands-mining-mineral-processing (Date accessed 16th August 2012): http://www.baseresources.com.au
3. Davies T.C and Osano O. (2005): Sustainable mineral development case study from Kenya *Geological Society, London, Special Publications* p. 87-93.
4. Government of Kenya (1940): The Mining Act (*Cap 306*), Government Press, Nairobi.
5. Government of Kenya (2000): Environmental Management and Coordination Act. Government Press, Nairobi.
6. Horkel A.D., Neubauer W., Niedermayr G., Okelo R.E., Wachira J.R., and Werneck W.(1984): Notes on the Geology and Mineral Resources of the Southern Kenyan Coast. Kenya-Austria mineral exploration project Rpt III, 106pp.
7. Kenya Flouspar Company (2012): mining operation and processing of ore information, (Date accessed 13th August 2012): www.kenyafluorspar.com
8. Kenya National Bureau of Statistics, (2012): Kenya facts and figures for 2012, (Date Accessed 13th August 2012) http://knbs.or.ke/downloads/pdf/Kenyafacts2012.pdf
9. Mason J.E and Theuri F.G. (1980): Industrial mineral development in Kenya. 4th Industrial Minerals Intern. Congress, Atlanta, USA 1980, 18pp.
10. Mathu E.M and Davies T.C. (1996): Geology and the Environment in Kenya; *Journal of African Earth Sciences*, vol.23, No.4

11. Mines and geological department Nairobi (1962). Geological map of Kenya, published by Survey of Kenya.
12. Ministry of environment and Mineral Resources (2010). The National Mineral and Mining Policy.
13. Straaten V. P. (2002). Rocks for Crops: Agrominerals of sub-Saharan Africa. ICRAF, Nairobi, Kenya, 338 pp.
14. Theuri F.G. (1994). The Kenyan industrial minerals sector. In: Mathers, S.J and Notholt, A.J.G (eds.) Industrial Minerals in Developing Countries. AGID Geosciences in International Development 18:167-172.
15. Walsh. (1970). The Geology and Mineral Resources of Kenya report.

Chapter 7

Bench Scale Feldspar Flotation for Malaysian Pegmatite Ore Using Non Hydrofluoric Modifier and Mixed Collector

Ismail Ibrahim and Md Muzayin Alimon

Mineral Research Centre, Minerals and Geoscience Department Malaysia
Jalan Sultan Azlan Shah, 31400 Ipoh, Perak, Malaysia
E-mail: is.ibrahim@jmg.gov.my

ABSTRACT

A bench scale feldspar flotation study by using sulfuric acid (H_2SO_4) as a modifier, a mixed type collector of N-tallow 1, 3–propanediamine-dioleate (duomeen TDO) and Aero 65 as a frother, was carried out. In this study a Malaysian pegmatite sample where the major components are feldspar and quartz, was used. Based on the field emission scanning electron microscope (FESEM) micrograph on the sample, it was indicated that fine quartz particles were found to be disseminated in the feldspar. The behaviour of duomeen TDO in the feldspar–quartz flotation system has been investigated through zeta potential and fourier transform infrared (FTIR) measurements. Several series of feldspar flotation tests were carried out using different set of flotation parameters. The parameters involved are pH of the slurry, particle size range of the feed material and different reagent (duomeen TDO, H_2SO_4 and Aero 65) dosages. The results show that high feldspar grade concentrate (86.79 per cent) and a good recovery (87.96 per cent) can be obtained at pH 2 by using 800 g/t duomeen TDO and 90 g/t Aero 65.

Keywords: *Feldspar, Quartz, Bench scale flotation, Flotation, Duomeen TDO, Non-hydrofluoric modifier.*

Introduction

Feldspar is a basic raw material with a significant proportion of Na-feldspar (albite, $NaAlSi_3O_8$) and K feldspar ($KAlSi_3O_8$, microcline or orthoclase) which exist in the same matrix. The separation of feldspar minerals by flotation is challenging due to their identical crystalline structures and close similarities in their physical and chemical properties (Rao and Forssberg, 1995; Demir, *et al.*, 2003). The basic feldspar structure consists of a ring of four tetrahedral unit; potassium and sodium feldspar have three silicon tetrahedral and one aluminium tetrahedron, while in calcium feldspar half of four tetrahedral units are silicon-based and another half are aluminium-based (Smith, 1974 and Graveling, *et al.*, 1997). Silica is often associated with feldspar, as quartz in pegmatite and silica sand in feldspathic deposit. The basic structure of quartz (SiO_2) consists of tetrahedral with oxygen anions as it apexes and a centrally positioned silicon atom. A complete tetrahedron, then, has four negative ions (Huang *et al.*, 2001).

When feldspar and quartz are ground, the surfaces which consist of silicon and aluminium atoms (for feldspar) and silicon (for quartz) are unsaturated due to anionic tetrahedron breakage. In a dry environment, broken Si-O bonds will result in reactive surface sites, on which atmospheric water molecules can chemisorb to cover the surface with hydroxyl group (silanol group). However when placed in water, the surface atoms are saturated with hydroxyl group and thus, the dissociation of surface hydroxyl H^+ or OH^- at low and high pH, respectively, will create either a positively or negatively charged surface (Rao *et al.*, 1995; Graveling *et al.*, 1997; Alkan *et al.*, 2005 and Karaguzel *et al.*, 2005). The significant isoelectric point is the pH at which the zeta potential is zero where the density of positive and negative sites is equal (Franks and Meagher, 2003).

The flotation of feldspar from quartz is known as the most suitable method for the separation process. In a conventional process, feldspar is separated from quartz using a cationic amine collector and hydrofluoric (HF) acid as an activator for feldspar. However, as the waste from this process may pollute the working environment and can create health problem to the operators, and a high cost is required in order to keep the process as environmental friendly as possible (Oteyaka, *et al.*, 2000; Viyadhar and Rao 2002). The environmental impacts may be from the potential fluoride ion contamination of both surface and subsurface water sources (Sadowski, *et al.*, 2008). Mixed cationic and anionic collectors by means of duomeen TDO such as 1,3 propylene diamine-dioleate (Malghan, 1976 and Malghan, 1981) and 1,3 propane-diamine-dioleate (Viyadhar *et al.*, 2002; Viyadhar and Rao, 2007) were found to be effective for feldspar flotation without using HF acid. Several tests on single mineral flotation without using HF acid were carried out in order to obtain high recoveries of feldspar (Tang, 1993; Rao and Forsberg, 1993; Liu, *et al.*, 1993; Salmavy, El M.S., *et al.*, 2000). A HF^- free flotation process for recovering feldspar by using a collector system comprising a mixed cationic/non-ionic surfactant has also been carried out. It was found that feldspar may be recovered in high yields and high selectivity, including one or more higher aliphatic amines and one or more higher alcohols (Filippov, Lev O., 2011).

This paper aims at evaluating feldspar behaviour and its flotation process performance using mixed cationic and anionic collector N-tallow 1,3 propane diamine-dioleate (duomeen TDO). The flotation process performance was measured using Denver flotation cell. In this study, characterizations of sample such as chemical and physical properties which may affect the performance were also being noted.

Materials and Methods

Sample and Sample Preparation

Sample of pegmatite rock used in this experimental works was obtained from Bukit Mor Quarry, Johore, Malaysia. First, the rock samples were crushed using crushers and mixed thoroughly. Then, grindings of crushed materials were performed for 17 minutes in a laboratory ball mill with 10 kg of mixed steel balls (5 kg of 5 cm, 3 kg of 3.75 cm and 2 kg of 2.5 cm diameter) to achieve a particle size of d_{80} = 600 µm, in which this particle size corresponds to the liberation size as confirmed by mineralogical analysis, and followed by sieving. Then, the sieved materials (-600+45 µm) from batches were mixed together and was passed through wet high intensity magnetic separator (WHIMS) to get rid of iron content. Flotation tests using bench scale Denver cell on those non-magnetic materials were then performed. Sample Characterizations by X-ray diffraction (XRD), field emission scanning electron microscope (FESEM), energy dispersive X-ray (EDX) and the Fe content (analyzed by atomic absorption spectrophotometer, AAS) of the feed material for the flotation tests were carried out to support the studies.

In order to study the behaviour of the selected collector (duomeen TDO) on feldspar and quartz in the flotation system, free feldspar and quartz were manually selected, ground and followed by sieving to produce particle size fraction of -150 + 38 µm. The grinding process was repeated until the sufficient sample was obtained. Some of the particles were further pulverized using agate mortar and sieved to -5 µm using ultrasonic bath. This fraction was then analyzed by zeta potential and FTIR. For feldspar, a single mineral flotation tests using Hallimond tube were also performed.

Chemical Reagent

(i) A mixed cationic-anionic diamine collector of N-Tallow 1,3 propane diaminedioleate (duomeen TDO) that was obtained from Akzo Nobel AB, Sweden was used in the experimental works. Duomeen TDO is a compound consists of reacting one molecule of alkyl (tallow) propane diamine [R-NH-CH_2-CH_2-CH_2-NH_2] and two molecules of oleic acid [$CH_3(CH_2)_7CH=CH-(CH_2)_7COOH$]. Prior flotation process, duomeen TDO collector was dissolved in distilled water. Analytical grade of sodium hydroxide was used as pH adjustment. (ii) Diluted Aero 65 was used as frother. (iii) Sulfuric acid.

Sample Characterization

X-ray diffraction (XRD), field emission scanning electron microscope (FESEM) and energy dispersive X-ray (EDX) were used to support the studies. The Fe content was analyzed by atomic absorption spectrophotometer (AAS).

Zeta Potential

Zeta potential was also carried out in order to obtain an indication of the magnitude of the potential at the beginning of the diffuse double layer around the particles. Electrokinetic properties such as potential determining ions (pdi) and isoelectric point (iep) in an aqueous of fine particles are very important factors in understanding the adsorption mechanism of organic and inorganic species at the solid and solution interface. To perform the zeta potential test, feldspar and quartz particles -5 µm were used at a concentration of 0.025 g/100 ml. Deionised water with constant ionic strengths of 10^{-3} M was prepared by adding with KNO_3 with particular pH. The solution was then added with the mineral (0.025 g/100 ml) and conditioning for 30 minutes Solutions in the range of pH 2 to pH 10 were prepared and then the mineral was added. After 30 minutes conditioning, the pH of the suspension was recorded again and regarded as the pH of the measurement. Zeta potential was determined using Brookhaven Zeta plus zeta meter. The softwares allow the direct reading of zeta potential calculated from the electrophoretic velocity which is proportional to the electric field, with the proportionality constant called the electrophoretic mobility using new approach (Bruce *et al.*, 1993; Fadyen and Fairhurst, 1993).

FTIR Measurement

Fourier transformed infrared (FTIR) spectroscopy is a measurement technique for collecting infrared spectra and hence the chemical bonds in a molecule can be determined. Since the collector used was TDO Duomeen, the relative adsorption of this collector on feldspar and quartz were accessed by the intensity of the adsorption band. Thus the alkyl band characteristics, v_{as} (CH_3), v_{as} (CH_2) and $v_s(CH_2)$ as well as amine spectrum due to v_{as} (NH_2) and v_s (NH_2) could be determined. At the same time, the area under the alkyl chain can be measured. To perform the FTIR test, 0.1 g of the mineral powder (-5 µm) was conditioned in an Erlenmeyer flask containing 100 ml collector solution at specified pH and concentration for 60 minutes (Vidyadhar *et al.*, 2007). After equilibration, the suspension was filtered through millipore filter paper and the solids were air-dried overnight at room temperature. The FTIR spectras were obtained with Perkin Elmer Spectrum One, an average of 400 scans measured at 4 cm-1 resolution with narrow band. The samples were mixed with KBr matrix.

Single Mineral Flotation

The Hallimond tube flotation tests for feldspar and quartz at pH 2 with different dosage of collector were performed. The performance of the flotation was measured based on the recovery of a single mineral. pH was controlled by sulfuric acid (H_2SO_4) and NaOH. Prior to flotation, 250 ml of collector solution was prepared in a beaker and agititate by magnetic stirrer. The solution was then transfered into the Hallimond Tube and 1 gram of the mineral was then added. The suspension was conditioned for five minutes followed by flotation for 1 minute at a constant air flowrate of 10 ml/min. When flotation time was done, the air flow and stirrer was stopped. Float product was collected, filtered, dried and weighed. The weight of product obtained will be divided by 1 g and multiplied by 100 to get per cent of recovery.

Bench Scale Flotation Cell

Flotation tests were carried out in a laboratory scale Denver flotation machine with a cell capacity of 2.5 litre. Initially for every test, 500 grams of ground non-magnetic materials was run in 2 liters of pulp. The experimental work was conducted at constant parameters of pH controlled by using H_2SO_4, 20 per cent solids, 90 g/t frother (Aero 65) and by varying the parameters of duomeen TDO dosage, pH and size as shown in Table 7.1.

Table 7.1: Series of Experimental Works for Feldspar Flotation on Non-Magnetic Fraction using Duomeen TDO

Series No.	Size (μm)	pH	Per cent Solids	pH Modifier	Duomeen TDO g/t	Aero 65 (g/t)
1	–600+45	3	20	H_2SO_4	200,300,400, 500,600,700	90
2	–600+45	2	20	H_2SO_4	200,300,400, 500,600,700	90
3	–212+45, –300+45, –425+45, –500+45, –600+45	2	20	H_2SO_4	600	90
4	–300+45	2	20	H_2SO_4	400,600,700, 800,1000	90

Results and Discussion

Sample Characterization

The XRF results of pegmatite samples and individual feldspar are shown in Table 7.2. It can be seen that pegmatite sample contains high content of 73.01 per cent of SiO_2 and 15.84 per cent of Al_2O_3, other than that were Fe_2O_3, Na_2O and K_2O were 0.81 per cent, 3.02 per cent and 6.01 per cent respectively. From the calculation by feldspar convention technique, it was found that the pegmatite sample contains 61.07 per cent of feldspar and 27.33 per cent of quartz. XRF analysis on feldspar sample which was collected manually from the crushed particles, shows that the composition of SiO_2, Al_2O_3, Fe_2O_3, Na_2O and K_2O were 68.50 per cent, 17.48 per cent, 0.11 per cent, 2.66 per cent and 11.21 per cent respectively. Using similar calculation method indicated that the feldspar sample contain approximately 89 per cent of feldspar and 9 per cent quartz. It can also be seen that the percentage of potassium (K)

Table 7.2: The Chemical Composition of Pegmatite and Feldspar Sample

Sample	SiO_2 (%)	Al_2O_3 (%)	Fe_2O_3 (%)	MgO (%)	CaO (%)	Na_2O (%)	K_2O (%)	TiO_2 (%)	LO (%)
Pegmatite	73.01	15.84	0.81	0.11	0.75	3.02	6.01	0.02	0.54
Feldspar	68.50	17.48	0.11	0.01	0.42	2.66	11.20	0.03	0.51

was very much higher than percentage of sodium (Na), therefore feldspar sample can be classified as K feldspar.

Results for XRD analyses on pegmatite and individual feldspar samples are shown in Figures 7.1(a) and 7.1(b) respectively. The analyses were based on strong reflection at 2θ angles. Figure 7.1(a) illustrates that three main phases were identified. These were microcline, albite and quartz with the formula $KAlSiO_3O_8$, $NaAlSi_3O_8$ and SiO_2 respectively. The patterns of x-ray diffraction of individual feldspar as shown in Figure 7.1(b) revealed that similar minerals present in this sample. Both samples showed that the feldspar can be categorized as combination of potassium (potash) and sodium feldspar.

Figure 7.2(a) shows the FESEM micrograph on polish section sample of pegmatite rock. The presence of Fe originated from locked iron bearing mineral. The disseminations of iron particles interlocked at area A were traced by FESEM as shown in Figure 7.2(a) and the elements presence in area A were confirmed by EDAX in Figure 7.2(b) shows the presence of O (44.42 per cent), Fe (23.64 per cent), Na (1.58 per cent), Al (15.16 per cent), Si (12.70 per cent) and Mg (2.51 per cent). The EDAX analysis on feldspar at area B was also run and revealed the presence of O (37.73 per cent), Na (5.93 per cent), Al (10.03 per cent), Si (28.22 per cent) and K (18.10 per cent), while for Area C in Figure 7.2(d) revealed the presence of Si (52.33 per cent) and O (45.11 per cent).

Figure 7.3(a) shows the FESEM micrograph on polish section sample of individual feldspar. It can be seen that the quartz particles identified a indefinite shapes and disseminated in feldspar approximately between 10 to 130 μm. The elemental compositions of the individual feldspar at Area A as measured by EDAX (Figure 7.3b) were identified as Al (9.52 per cent), Si (28.44 per cent), K (10.13 per cent), O (46.48 per cent) and Na (2.32 per cent), while interlocked quartz at Area B were Si (45.12 per cent) and O (52.32 per cent) as shown in Figure 7.3(b).

Zeta Potential Measurements

Figure 7.4(a) illustrates the effect of pH on zeta potential of feldspar and quartz. It can be seen that an increase of pH in suspension resulted in an increasing in the negative charge of both samples until pH 10. The results showed that the isoelectric point (iep) of quartz was at pH 2 while the extrapolation of the curves in the case of feldspar indicated iep of about pH 1.90. At pH 2, feldspar exhibited -7.89 mV whereas quartz approximately 0.00 mV. In general feldspar showed higher negative at all pH charge compared to quartz.

The zeta potential of feldspar and quartz as a function of duomeen TDO dosage at pH 2 and pH 3 is shown in Figure 7.4(b). Previously feldspar was negatively charged at pH 2 and pH 3 which was due to zeta potential attained with solely deionised water and then the doubly charged positive diamine species adsorbed through electrostatic interactions could be leading to its flotation. However, from the results obtained from both feldspar and quartz, pH 3 posed slightly higher of mV values compared pH 2. This was also due to earlier zeta potential attained with solely deionised water which gave higher mV value at pH 3 for feldspar (-19.93 mV)

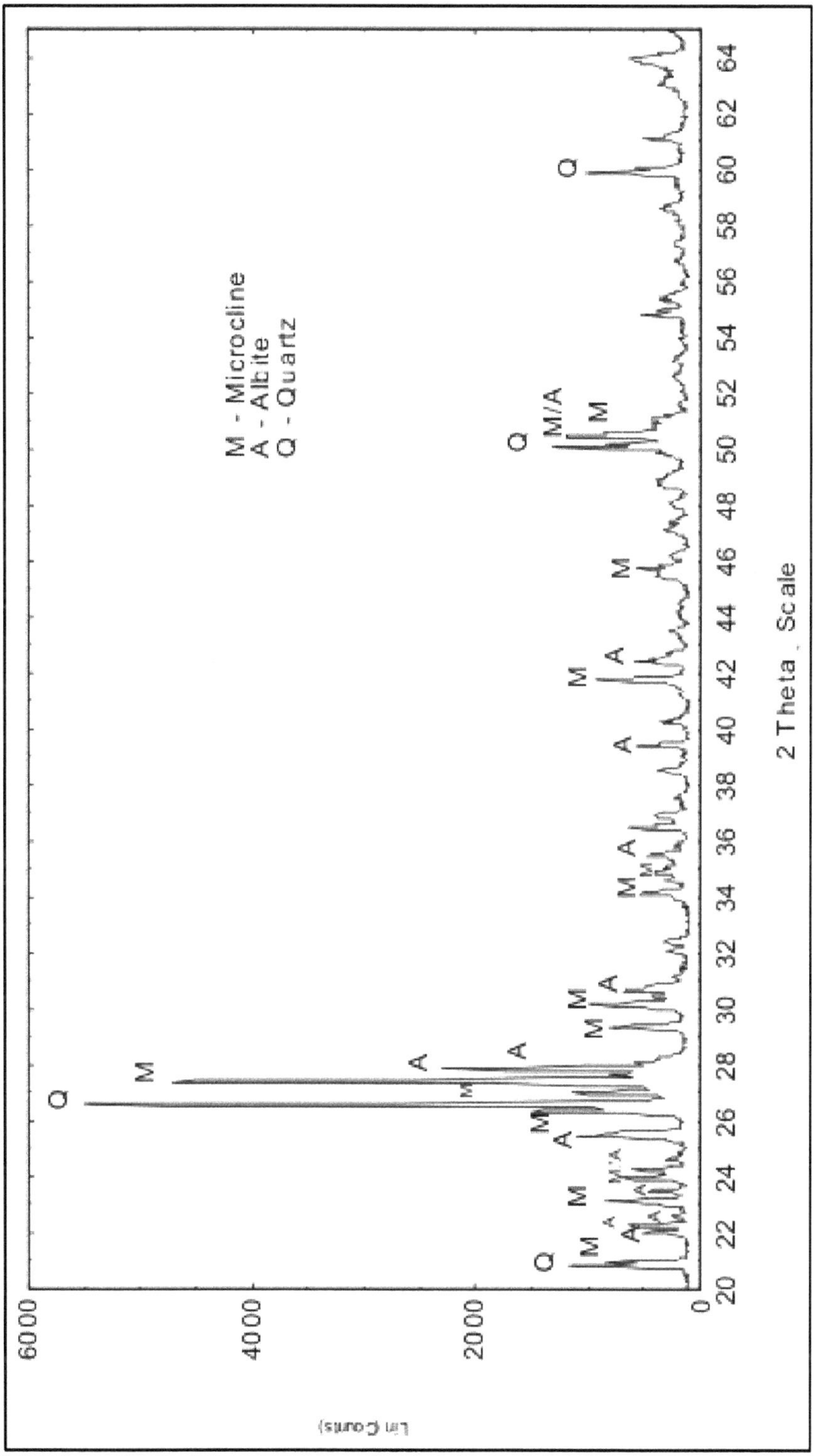

Figure 7.1(a): XRD Diffractogram of Pegmatite Sample Contains the Major Minerals of Microcline, Albite and Quartz.

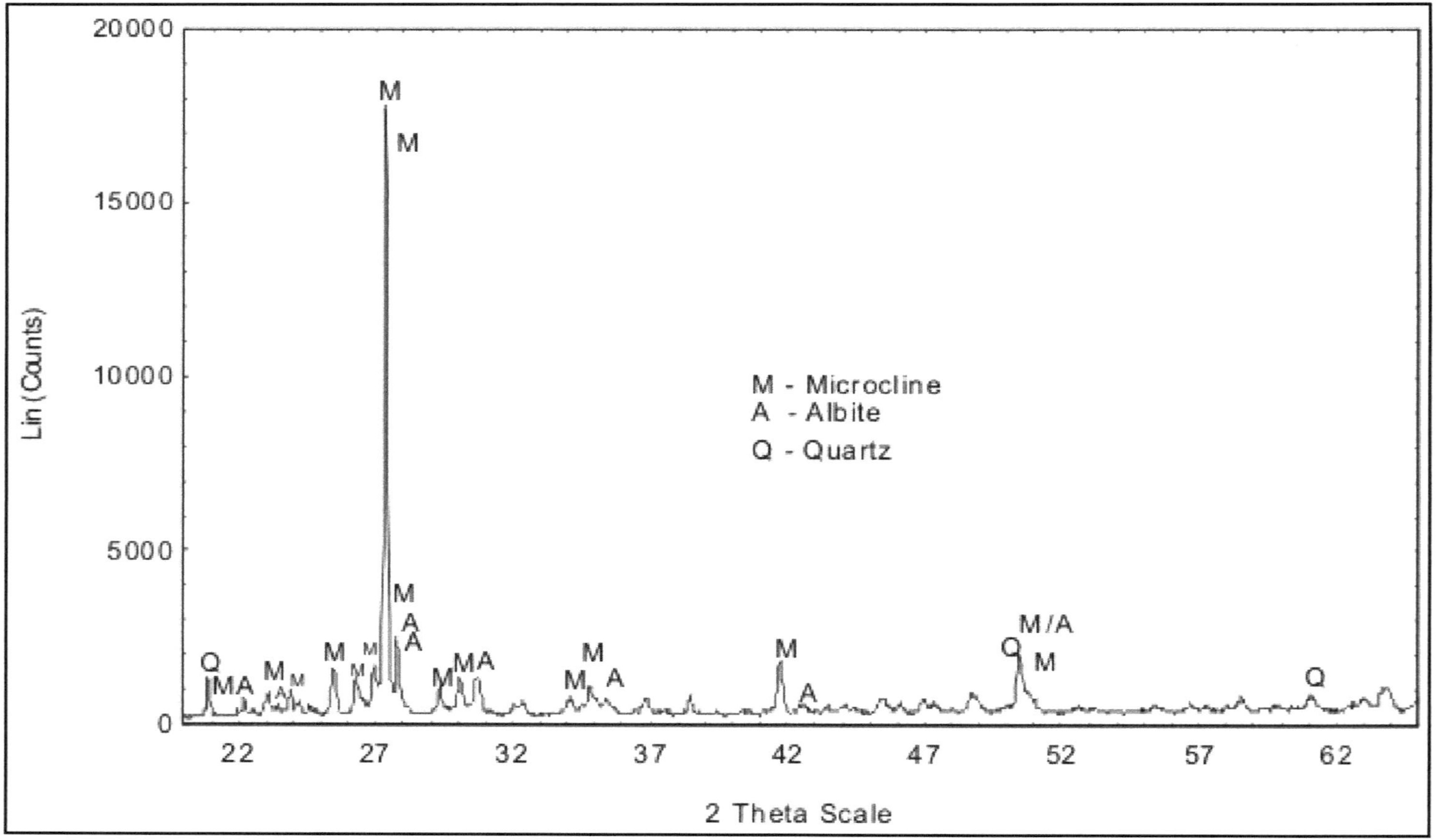

Figure 7.1(b): XRD Diffractogram has Identified that the Individual Feldspar Sample Contains the Major Minerals of Microcline, Albite and Quartz.

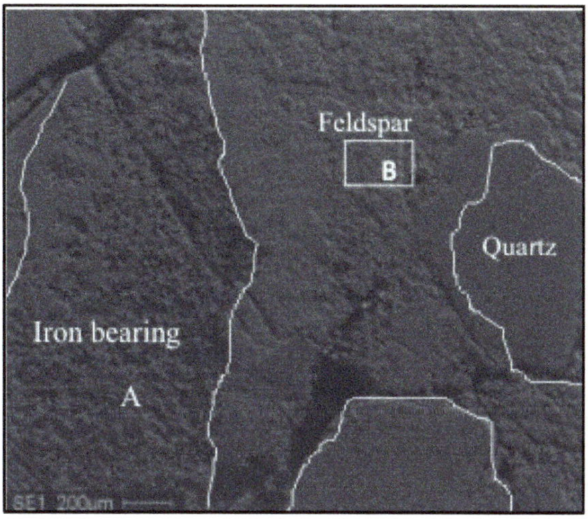

Figure 7.2(a): FESEM Micrograph Shows the Iron Bearing Mineral Interlocked in Feldspar in the Pegmatite Rock Sample.

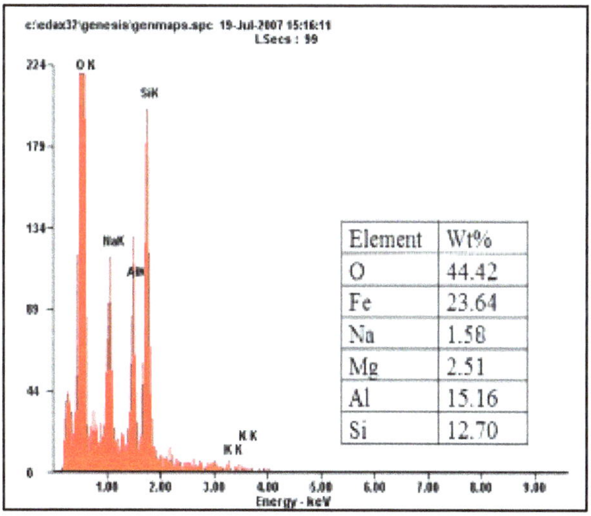

Figure 7.2(b): EDAX Spectrum at Location A from Figure 7.2(a).

and quartz (-6.57 mV) compared to pH 2 for feldspar (-7.89 mV) and quartz (0.0 mV). A large difference of zeta potential (around 26 mV) between feldspar and quartz was found in 60-80 ppm duomeen TDO resulted higher selectivity of feldspar can be obtained at pH 2. Since quartz posed around 13-15 mV between 60-80 ppm duomeen TDO, resulted quartz also would be partially floated if the actual feldspar ore is employed.

Since feldspar was negatively charged at pH 2, the doubly charged positive diamine species adsorbed through electrostatic interactions leading to its flotation but not adsorbed on quartz due to zero charge. The adsorption principally occurred

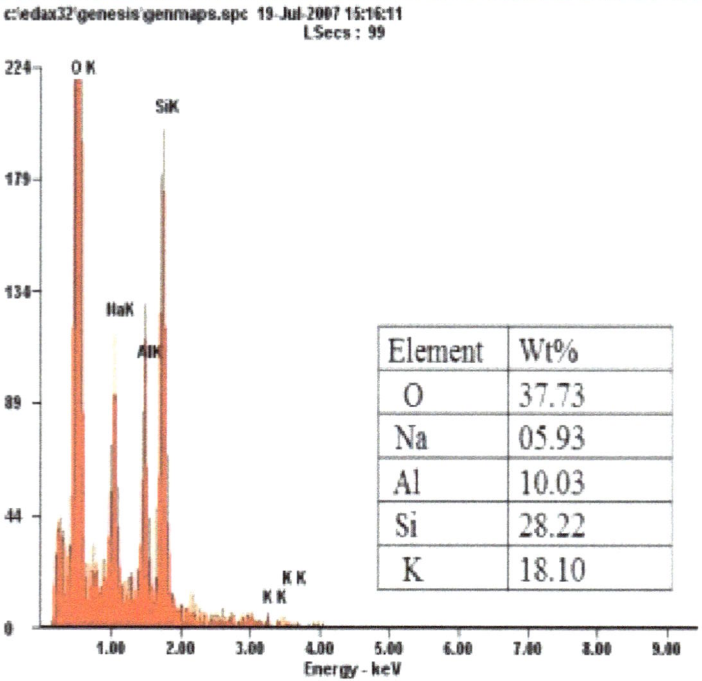

Element	Wt%
O	37.73
Na	05.93
Al	10.03
Si	28.22
K	18.10

Figure 7.2(c): EDAX Analysis on Feldspar at Area B from Figure 7.2(a).

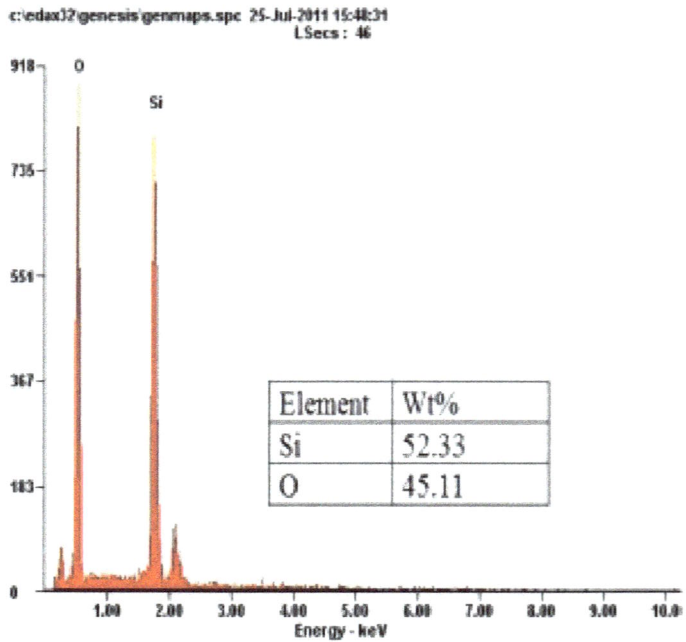

Element	Wt%
Si	52.33
O	45.11

Figure 7.2(d): EDAX Spectrum Shows Quartz Mineral at Area C.

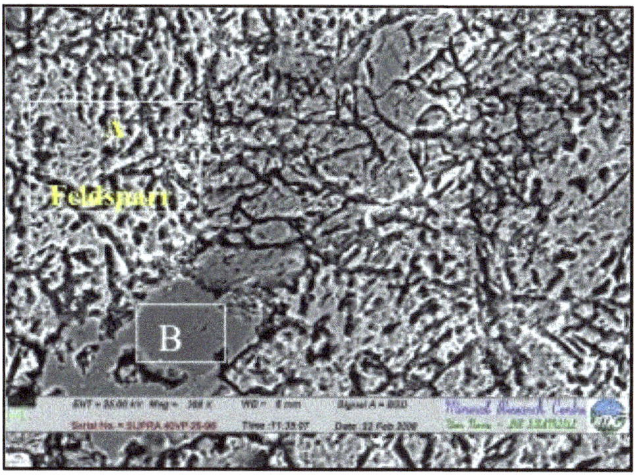

Figure 7.3(a): FESEM Micrograph Shows the Quartz Disseminated in Feldspar.

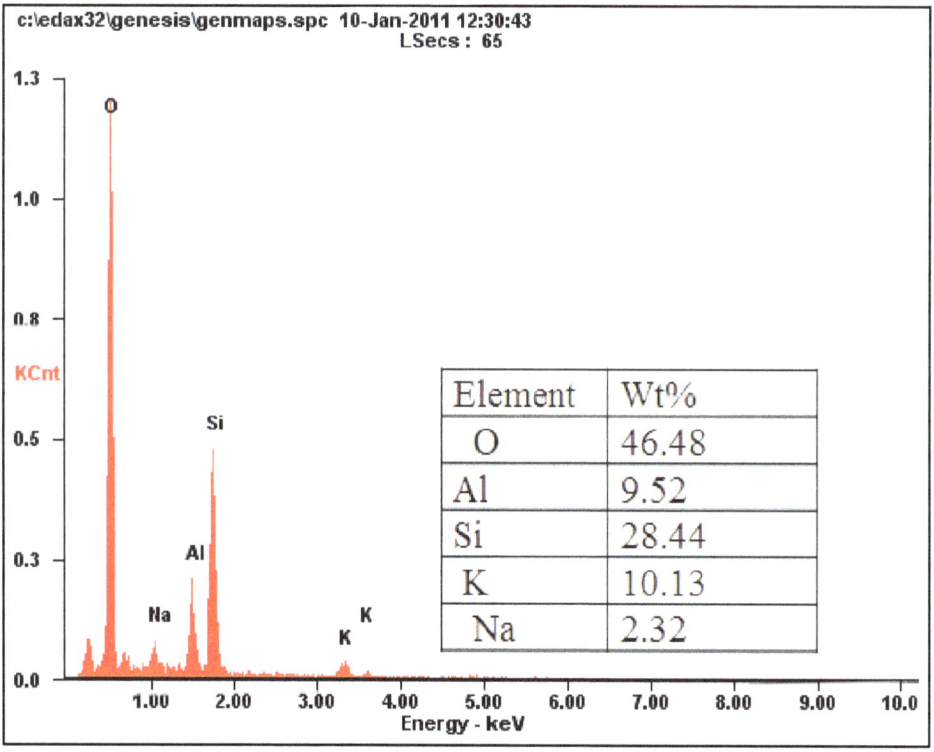

Figure 7.3(b): EDAX Spectrum Shows Feldspar Mineral at Area A.

was due to duomeen TDO which is a compound consists of reacting one molecule of alkyl (tallow) propane diamine [R-NH-CH_2-CH_2-CH_2-NH_2] and two molecules of oleic acid [$CH_3(CH_2)_7CH$=CH-$(CH_2)_7COOH$]. To yield di-oleate salt, the structural formula [R-NH_2-CH_2-CH_2-CH_2-$NH_3)^{++}$ $2(C_{17}H_{33}COO)^-$ indicates that the compound

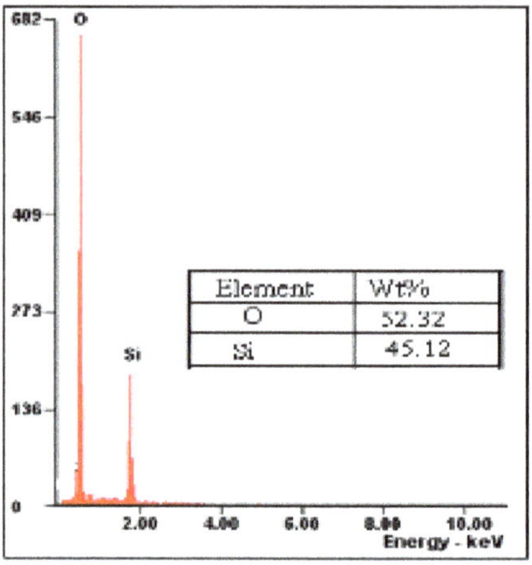

Figure 7.3(c): EDAX Spectrum Shows Quartz Mineral at Area B.

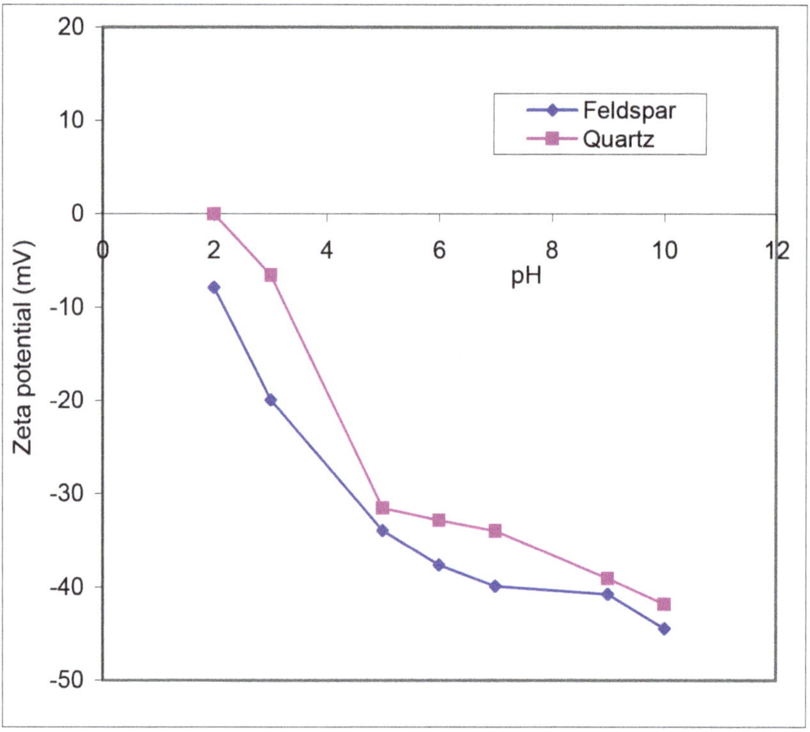

Figure 7.4(a): Effect of pH on Zeta Potential for Feldspar and Quartz.

has both primary and secondary group (Malghan, 1976). The diamine and oleate molecules forming the diamine-dioleate compound could be through to hydrogen bonding involving two electronegative atoms of O and N of the respective functional groups (Vidyadhar *et al.*, 2002).

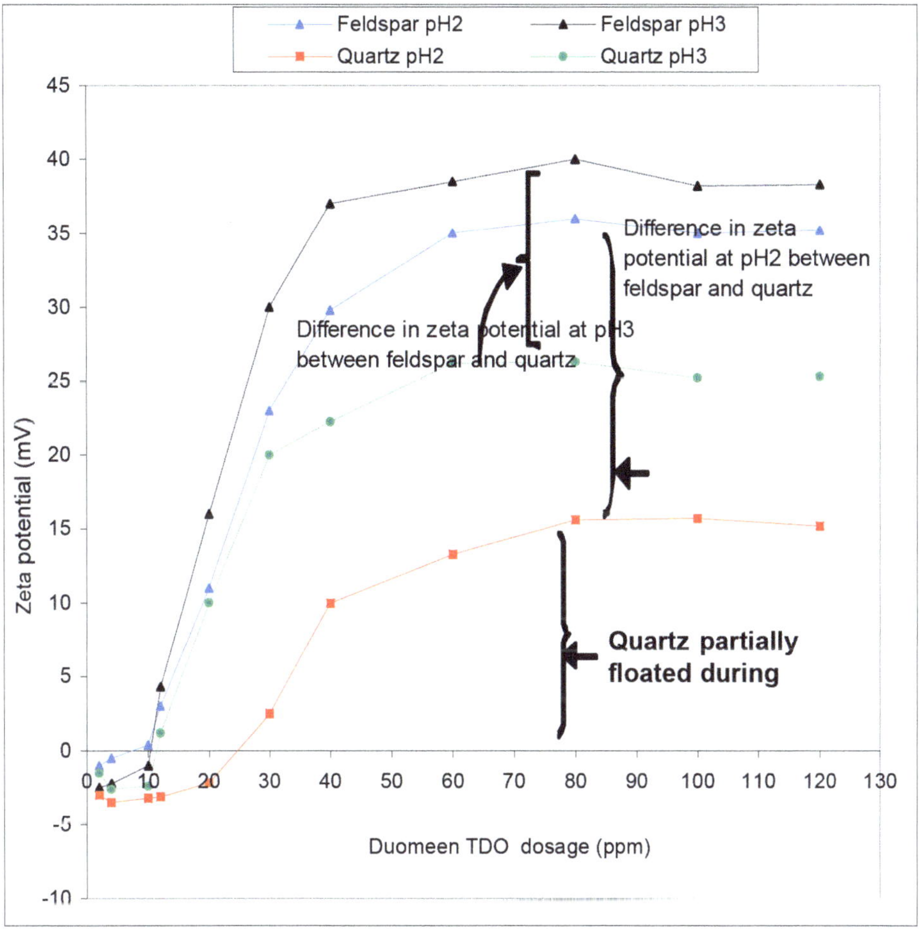

Figure 7.4(b): The Zeta Potential of Feldspar and Quartz as a Function of Duomeen TDO Dosage at pH2 and pH3.

$\equiv SiOH...H_2N\text{-}R \leftrightarrow \equiv SiO^- ...H_3^+N\text{-}R$

Where, negative charges (SiO^-) exist on the feldspar surfaces.

FTIR Measurement

To investigate the effect of duomeen TDO dosage on feldspar and quartz, two set of experiments were performed in which the pulp were subjected to pH 2. The adsorption plot of the peaks for feldspar and quartz are illustrated in Figures 7.5(a) and 7.5(b) respectively. The FTIR spectra of feldspar and quartz treated with 100 ppm duomeen TDO showed the bands characteristic of alkyl chains, υ_{as} (CH_3), υ_{as} (CH_2) and υ_s (CH_2) groups between region 3000-2800 cm^{-1}, at around 2955 cm^{-1}, 2926 and 2854 cm^{-1}. Figure 7.5(c) shows the peak area under the alkyl chain bands between region 3000-2800cm^{-1} was measured and plotted as a function of dosage The intensity

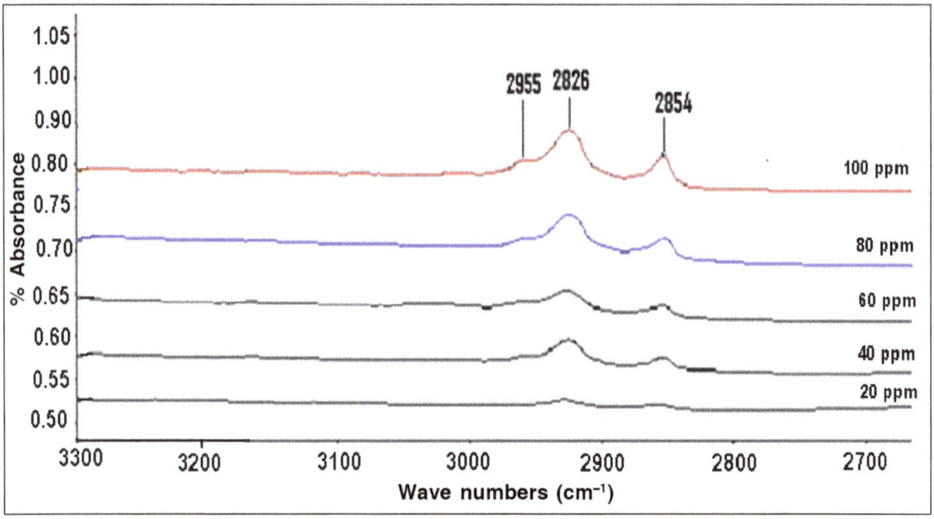

Figure 7.5(a): Adsorption of Duomeen TDO Dosage on Feldspar at different Dosages.

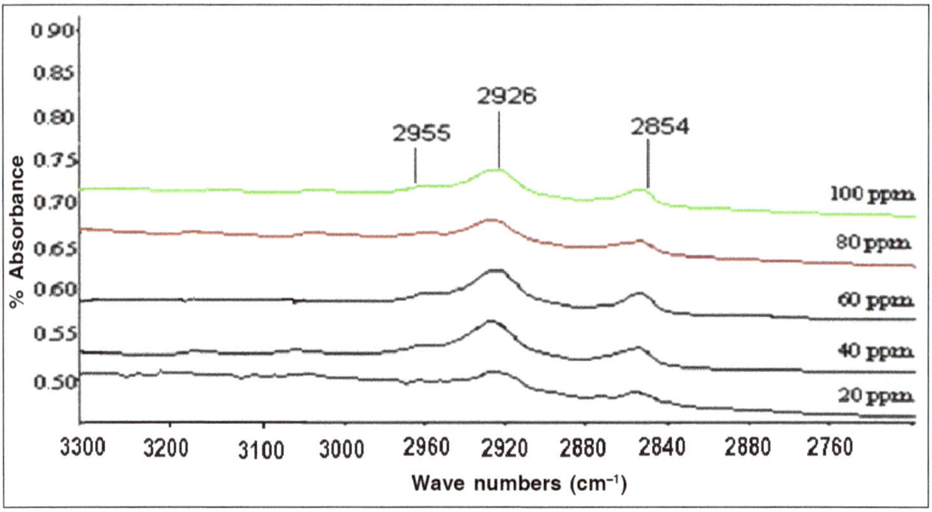

Figure 7.5(b): Adsorption of Duomeen TDO on Quartz at different Dosages.

of the alkyl chain bands increased slightly higher for feldspar compared to quartz with increasing initial duomeen TDO dosages, emphasising more collector adsorption on feldspar. It was assumed that since the zero potential charge (zpc) for quartz was found to be 2, the duomeen TDO adsorption at this pH on quartz was also lower.

Single Mineral Flotation

The flotation of feldspar and quartz response differently by adding different dosage of duomeen TDO at pH 2 with sulfuric acid as pH controller is shown in

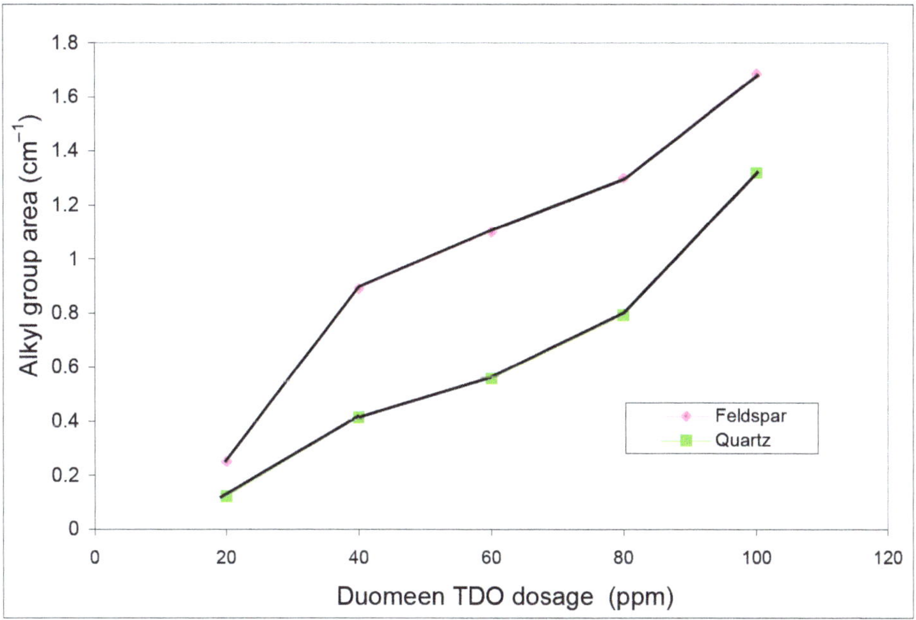

Figure 7.5(c): Area Under Alkyl Chain Bands (3000-2800 cm^{-1}) of Spectra of Feldspar and Quartz as a Function of Duomeen TDO Dosage.

Figure 7.6. In general the feldspar recovery increased with the increasing of duomeen TDO concentration. Feldspar floated more effectively compared to quartz which floated in lesser amount. Thus, the flotation of feldspar however started to occur at very low

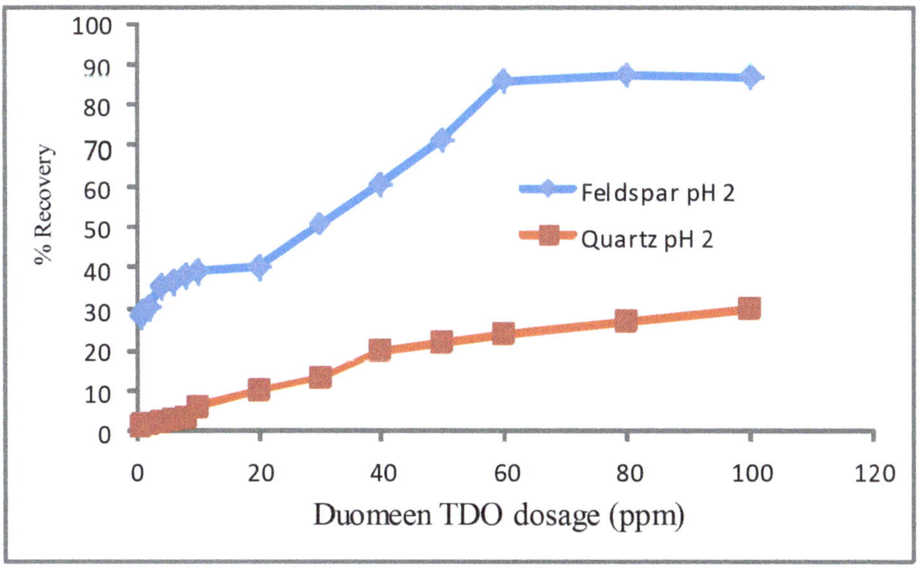

Figure 7.6: Recovery of Feldspar and Quartz as a Function of Duomeen TDO Dosage at pH 2.

concentration of duomeen TDO (0.8 ppm) with 28 per cent floated and increased gradually to 40.12 per cent at 20 ppm duomeen TDO, then its recovery rose to 86.0 per cent at 60 ppm duomeen TDO. Above this point the feldspar recovery remained similar. However, quartz recovery was seen to be much lower compared to feldspar.

Bench Scale Flotation

Effect of Duomeen TDO Dosage

The behavior of cationic and anionic collector mixtures in feldspar quartz flotation system was investigated through zeta potential. The results showed that the formation of molecular complex between oppositely charges collector ions through charge neutralization (Elworthy, *et al.*, 1968: Rao and Forssberg, 1993). The molecular complex was found to have higher surface activity than the individual components. The flotation response in using mixed cationic with anionic collector such as duomeen TDO were investigated through some different parameters. The reaction of feldspar flotation (series 1 and 2 from Table 7.1) as a function of duomeen TDO dosage from 200 g/t to 700 g/t at pH 3 and pH 2 is illustrated in Figure 7.7. For pH 3 it shows that feldspar grade slightly increased from 61.43 per cent at 200 g/t TDO duomeen to 70.05 per cent at 600 g/t and decreased to 57.96 per cent at 700 g/t duomeen TDO. However, feldspar recovery at pH 3 increased dramatically from 25.43 per cent (at 200 g/t) to 89.32 per cent at 700 g/t collector. Although feldspar recovery shown at

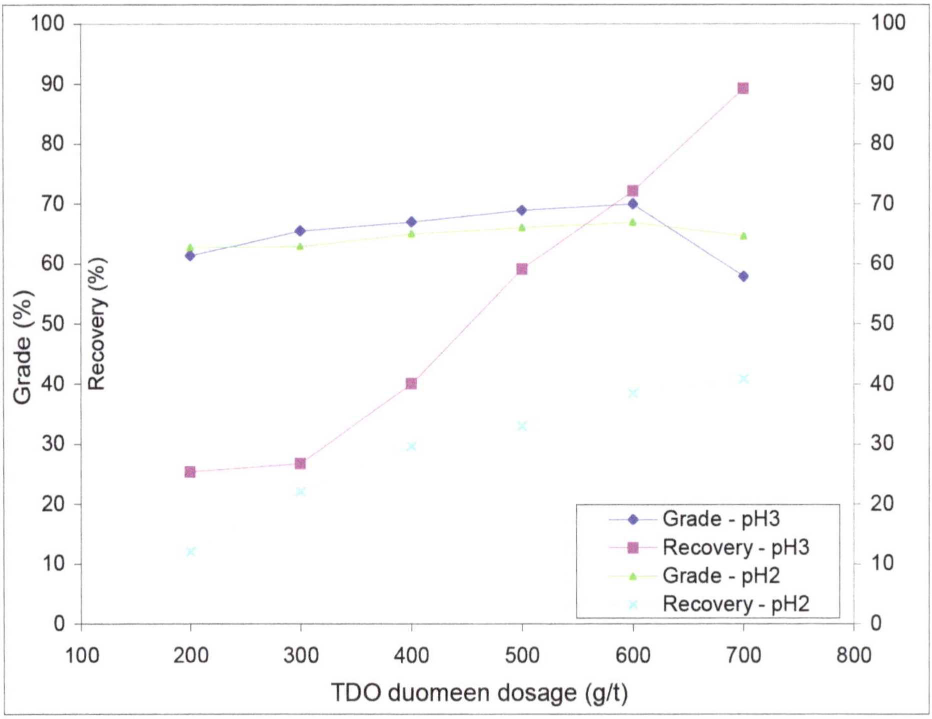

Figure 7.7: Effect of different Dosage of Duomeen TDO on Grade and Recovery of Feldspar at pH 2 and pH 3.

700 g/t of collector reached to 89.32 per cent, on the other hand its feldspar grade was 57.96 per cent due to quartz which was also reported to the froth. The results of feldspar grade and recovery for pH 2 present the same trend as shown by pH 3. However, the feldspar grade for pH 2 showed slightly lower (compared to pH 3) from 62.83 per cent at 200 g/t duomeen TDO to 67.10 per cent (600 g/t) and decreased to 64.78 per cent at 700g/t collector. The decreasing of feldspar grade at 700 g/t collector was due to increasing recovery of quartz.

Effect of Particles Size using Duomeen TDO

It is well known that that particle size undoubtedly imposes limitation upon flotation. Obviously the maximum size of particles that can be floated was limited by lifting power of the surface forces. Figure 7.8 present the feldspar grade and recovery as a function of size between (-212+45) µm to (-600+45) µm at pH 2 from Series 3 as shown in Table 7.1. The dosage of duomeen TDO used was constant at 600 g/t through out the tests. It can be seen that the feldspar grade increased from 70.10 per cent at -212+45 µm to 81.12 per cent at -300+45 µm, above this size the feldspar grade decreased gradually to 67.10 per cent. Meanwhile, the feldspar recovery was only increased from 35.29 per cent at (-212+45) µm to 61.10 per cent (-300+45 µm) and above this size the recovery continuously decreased to 38.51 per cent. The finer size fraction (-212+45 µm) gave than lower grade and recovery of feldspar was due to that fine particles, kinetically have low inertia and easily carried away by liquid streamlines around a bubble, thus giving low collision probabilities with the bubble. This phenomenon was reported by Somasundaran and Arbiter (1979) and Sivamohan (1990) mentioned that fine particles do not have sufficient kinetic energy to overcome

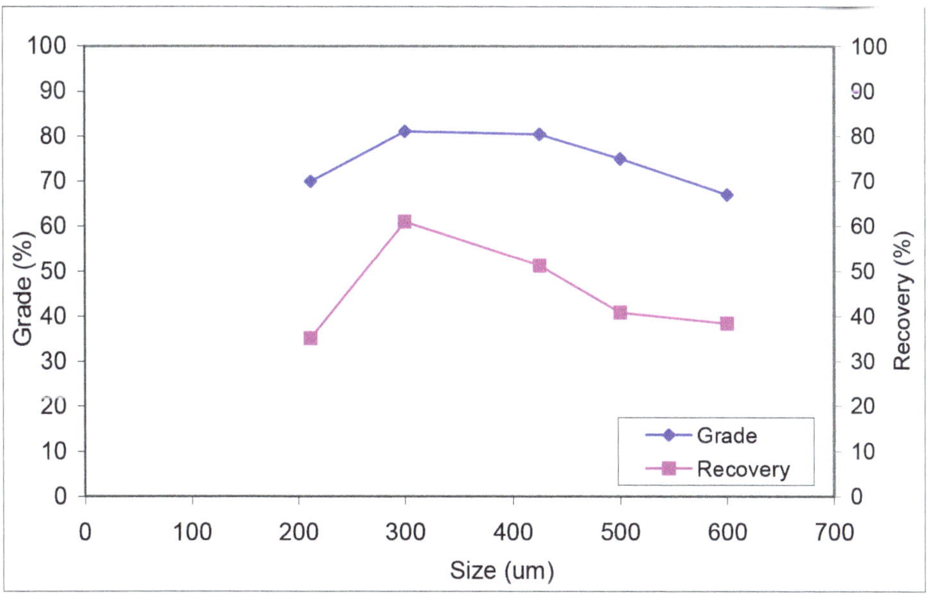

Figure 7.8: The Effect of Particle Size on the Feldspar Flotation at pH 2.

the energy barrier to rupture the liquid film between particles and bubbles to form three-phase contact.

Meanwhile the lower feldspar grade and recovery given by above 300 µm was due to that with the increase in the particle size the possibility of quartz presented on feldspar was more pronounced. This evident from FESEM micrograph has shown that the quartz disseminated on feldspar surface between 20 µm to about 1000 µm. At pH 2 (pzc of quartz=2), the existence of quartz would lead to depress the feldspar as it becomes its host materials along. Unlike if HF acid is used, it dissolved the quartz into solution occur by fluoride is subjected to be an exchange of silicon surface hydroxyl groups with fluoride. Hence it resulted in a reduction in the number of negative surface silicon sites to inhibit the adsorption activity of amine to silica surface. Secondly, as the size of particle increased eventually lead to feldspar particle exposure decreased, which reduced the adhesion between the particle and bubble, resulting in drop of mineral particle from froth as stated by Wills (1984). Since the maximum grade (81.12 per cent) was observed on (-300+45) µm sample. It may attributed to the increased in the availability of surface total feldspar for duomeen TDO adsorption as shown in Figure 7.8, which leads to an enhanced adsorption of duomeen TDO ions.

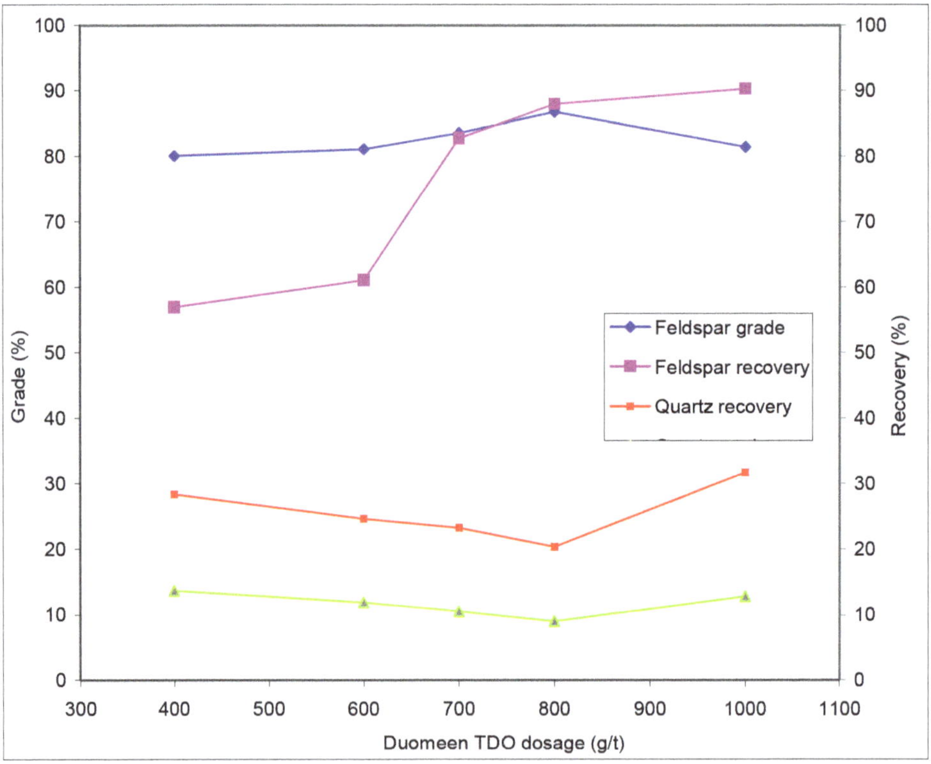

Figure 7.9: The Effect of Duomeen TDO Dosage on the Flotation of Feldspar at Size Fraction (–300+45) µm.

Effect of Duomeen TDO Dosage for –300+45 μm Fraction

The effect of duomeen TDO dosage was studied in the range of 400 g/t to 1000 g/t from Series 4 as shown in Table 7.1. The results of this experiment are shown in Figure 7.9. The feldspar grades were found slightly increased from 80.11 per cent (at 400 g/t) to 86.79 per cent at 800 g/t, after which the percent of grade decreased to 81.44 per cent at 1000 g/t duomeen TDO. The maximum grade (86.79 per cent) was observed at 800 g/t may attributed to the increased in the availability of surface total feldspar for duomeen TDO adsorption which leads to an enhanced adsorption of duomeen TDO ions. Above this dosage, a decreasing of feldspar grade was noticed could be due to formation of bilayer of duomeen TDO at the feldspar surface. However, the recovery of feldspar continuously increased from 57.02 per cent to 90.26 per cent at 400 g/t to 1000 g/t collector respectively. Even though the pzc of quartz was reported to be pH 2, quartz was found floated partially due to hydrogen bonding between silanol and diamine ion from duomeen TDO.

XRF Analyses of Flotation Concentrate

The XRF analyses of flotation concentrate by bench scale for all series are shown in Table 7.3.

Table 7.3: XRF Analyses of Flotation Concentrate by Bench Scale (Denver Cell)

Size (μm)	*pH*	*Duomeen TDO (g/t)*	*Aero 65 (g/t)*	*Feldspar Grade (%)*	*Feldspar Recovery (%)*	*SiO_2 (%)*	*Al_2O_3 (%)*	*Na_2O (%)*	*K_2O (%)*	*Fe_2O_3 (%)*
-600+45	3	200	90	61.43	25.43	73.61	14.94	3.1	6.10	0.28
		300		65.6	26.78	71.50	15.77	3.45	6.16	0.55
		400		67.09	40.08	72.80	15.74	3.60	6.20	0.38
		500		69.00	59.21	72.30	15.00	3.70	6.50	0.30
		600		70.05	72.19	73.67	14.92	3.70	6.20	0.29
		700		57.96	89.32	73.19	17.22	2.96	5.57	0.17
-600+45	2	200	90	62.83	12.09	67.07	22.49	3.13	6.15	0.53
		300		63.03	26.07	68.97	20.75	3.11	6.20	0.41
		400		65.11	29.67	68.39	20.97	3.27	6.35	0.39
		500		66.17	33.27	69.04	20.61	3.40	6.33	0.30
		600		67.10	38.51	68.60	21.06	3.44	6.43	0.32
		700		64.78	41.00	69.86	19.92	3.32	6.21	0.26
-212+45	2	600	90	70.00	35.29	70.0	15.89	3.88	6.30	52.8
-300+45				81.12	61.10	68.50	17.50	4.35	7.50	0.25
-425+45				80.49	51.34	68.55	16.54	4.33	7.42	0.26
-500+45				75.00	40.90	69.00	16.57	4.20	6.70	0.29
-600+45				67.10	38.51	68.60	16.5	3.44	6.43	0.32

Contd...

Table 7.3–Contd...

Size (μm)	pH	Duomeen TDO (g/t)	Aero 65 (g/t)	Feldspar Grade (%)	Feldspar Recovery (%)	SiO_2 (%)	Al_2O_3 (%)	Na_2O (%)	K_2O (%)	Fe_2O_3 (%)
-300+45	2	400	90	80.11	57.02	68.74	16.53	4.31	7.40	0.27
		600		81.12	61.1	68.50	17.50	4.35	7.50	0.25
		700		83.56	82.79	68.22	17.54	4.38	7.87	0.24
		800		86.79	87.96	68.20	17.56	4.68	7.99	0.25
		1000		81.44	90.26	68.98	17.00	4.32	7.60	0.1

Conclusions

Zeta potential results showed that the isoelectric points of feldspar and quartz were at pH 1.9 and pH 2.0, respectively. It is also shows that at pH 2, the magnitude of mV for feldspar was at -7.89 mV which can be considered higher compared to quartz (at 0 mV). In addition, a large difference of zeta potential between feldspar and quartz indicating higher selectivity of feldspar to the collector compared to quartz. For this reason, pH 2 was selected for the flotation test using Denver cell. From the overall study and observation, FESEM micrograph indicated and confirmed that mineral quartz with size range between 20 μm to about 1000 μm existed on feldspar surface. The existence of mineral quartz would lead to interfere as well as to depress its host material during feldspar flotation. Therefore, it is recommended best to have finer particle size of feldspar in order to liberate it from quartz and hence to generate better flotation results. Finally, at pH 2 by using 800 g/t of duomeen TDO and 90 g/t of Aero 65, the highest grade of feldspar (86.79 per cent) and good recovery (87.96 per cent) can be obtained through this complete study.

Acknowledgements

The authors would like to thank and also pleased to acknowledge the support:

(i) Director of Mineral Research Centre, Minerals and Geoscience Department and fellow staffs of the Centre especially to the staffs of Section of Mineral Processing Technology for their assistance of this research. (ii) Universiti Sains Malaysia Engineering Campus (iii) Lulea University, Sweden for technical assistance in the research study. (iv) Ministry of Natural Resources and Environmental, Malaysia for providing the financial.

References

1. Alkan, M., Demiirbas, Ö. and Dogan, M. 2005,"Electrokinetic properties of kaolinite in mono and multivalent electrolyte solutions." *Microporous and Mesoporous Materials*, 83, pp. 51-59.
2. Bruce, B. W., Walter, W. T. and David, F., 1993, "Zeta potential: A new approach." *Canadian Mineral Analysts Meeting*, Winnipeg, Manitoba, Canada.

3. Chernyshova, I. V., Rao, K. H., Vidyadhar, A. and Langmuir, A. V. S., 2000, "Mecahnism of adsorption of long chain alkylamines on silicates." *A spectroscopic study 1*, Quartz Langmuir, 16, pp. 8071-8084.

4. Demir, C., Bentli, I., Gulgönul, I. &Celik, M. S., 2003, "Effect of bivalent salts on the flotation separation of na-feldspar from k-feldspar." *Minerals Engineering*, 16.

5. Elworthy, P. H., Florence, A. T. and Farlane, C. B. M., 1968, "Solubilization by surface-active agents." *Ed.Chapman and Hall Ltd. London*, London.

6. Fadyen, P. M. and Fairhurst, D., 1993,"Zeta potentials of nano-ceramic material measurement and interpretation." *Canadian Mineral Analyst Meeting*, Winipeg, Manitoba, Canada.

7. Filippov, Lev O. (FR), 2011, "Flotation process for recovering feldspar from a feldspar ore." *European Patent Application*, Application number: 10290006.5.

8. Franks, G. V. and Meagher, L., 2003,"The isoelectric points of sapphire crystals and alpha alumina powder." *Colloids and Surfaces A, Physico-chemical and Engineering Aspects*, Volume 214, pp. 99-110..

9. Graveling, G. J., Ragnarsdottir, K. V., Allen, G. C., Eastman, J., Brady, P. V., Balsley, S. D. &Skuse, D. R., 1997, "Controls of polyacrylamide adsorption on quartz, kaolinite and feldspar." *GeochemicaetcosmochimicaActa*, 61, pp. 3515-3523.

10. Huang, P. and Fuerstenau, D. W., 2001," The effect of the adsorption of lead and cadmium ions on the interfacial behavior of quartz and talc." *Colloids and Surfaces A: Physicochemical and Engineering Aspects*, vol.177, 2-3, pp. 147-156.

11. Karaguzel, C., Can, M. F., Sonmez, E. &Celik, M. S., 2005, "Effect of electrolyte on surface free energy components of feldspar minerals using thin-layer wicking method." *Journal of colloid and interface science*, vol.285, no 1, pp. 192-200.

12. Liu, Y., Gong, H., Qiu, J. and Zhang, K., 1993, Aust. Inst. Min. Metall., Publ. Ser 3/93, Proc. XVIII *Int. Miner. Processing*, Parkville, Australia, pp. 857.

13. Malghan, S.G.,1976, "Selective flotation separation of feldspar in non fluoride medium." *SME Fall Meeting*, Denver, Colorado.

14. Malghan, S.G., 1981, "Effect of process variables m feldspar flotation using non-hydrofluoric acid system." *Mining Engineer*ing, pp. 1616-1622.

15. Oteyaka, B., Ucbas, Y., Bilir, K., Gursoy, H., Bozkurt, R., Ciftci, M., 2000, " Recovery of feldspar from altered granites." *In: Ozbayoglu, G., Hosten, C., Atalay, M.U., Hicyilmaz, C.*

16. Rao, H. K. and Forsberg, K. S. E., 1993, "Solution chemistry of mixed cationic/ anionic collectors and flotation separation of feldspar from quartz." Aust. Inst. Min. Metall., Publ. Ser 3/93, Proc. XVIII *Int. Miner., Sydney, Australia*, pp. 837-844.

17. Rao, H. K. and Forssberg, K. S. E., 1995, "Feldspar flotation: Theory and practice. In: Gash, A., Pradip, Kumar, R. (Eds), selected topic in *Mineral Processing*. Port City Press, Baltimore, pp. 86-117.

18. Salmavy, E.M.S., Kusaka, E. and Wakamatsu, T., 2000,"Adsorption mechanism of mixed cationicanionic collectors in feldspar-quartz flotation system." Elsevier Amsterdam, Proc. XXI *INT. Mineral Processing Congress*, pB8b-160.
19. Smith, J. V., 1974, "Feldspar minerals 3rd edition." USA Springer Verlag Berlin, New York, USA.
20. Sadowski,J., Sherrell, I. and Dobbins, M., 2008, "Separation of quartz and feldspar triboelectrostatic alternative."*Outotec, Physical Separation Technology*, External Newsletter, Issue 2 | December 2008.
21. Sivamohan, R., 1990, "The problem of recovering very fine particles in mineral processing, *Int.J.Miner.Process*, 28, 247.
22. Somasundaran, P. and Arbiter, N., 1979, "Beneficiation of minerals fines problems and research needs."*AIME*, Michigan, USA..
23. Tang, J., Sun, B., Cheng, Z., Chen, X., Zao, H. and Zhang, S., 1993, Proc. XVIII *Int. Miner. Processing Congress*, Parkville, Australia, pp. 851.
24. Vidyadhar, A., Rao, K. H. &Forssberg, K. S. E., 2002,"Adsorption of N tallow 1,3propanediamine - dioleate collector on albite and quartz minerals, and selective flotation of albite from greekstefania feldspar ore."*Journal of Colloid and Interface Science*, 248, 19-29.
25. Vidyadhar, A. &Rao, K.H., 2007, "Adsorption mechanism of mixed cationic/ anionic collectors in feldspar-quartz flotation system." *Journal of Colloid and Interface Science*, 306, pp. 195-204..
26. Wills, B. A., 1984, Mineral Processing Technology, 3rd edition, Pergamon press.

Chapter 8

Challenges of Mining and Processing Brine in Nigeria

E.C. Nwokemodo[1], M.O. Omojola[2] and A.P. Onwualu[3]

Raw Materials Research and Development Council,
Maitama, Abuja, Nigeria
E-mail: [1]ebelemodos@yahoo.com, [2]omojolamoses@hotmail.com,
[3]ponwualu@yahoo.com

ABSTRACT

Salt is an essential element in the diet of not only humans but of animals and plants. In addition, its industrial and other uses are numerous. Nigeria imports about 99 per cent of its annual salt requirement which is about 600,000 metric tonnes. The remaining 1 per cent is produced from brine occurrences in the country using indigenous method. The indigenous method involves concentration of brine through evaporation in open fire using firewood to packaging of the crude salt. The entire process produces about 12-15kg of salt in 28 days with attendant impurities in the form of calcium, magnesium and potassium salt as well as insoluble solids, thereby making it almost unfit for human consumption. The improved salt production method developed by the Raw Materials Research and Development Council in collaboration with Federal Polytechnic, Nasarawa, Nigeria produces about 56kg of salt in 8 hours with minimal impurities through the use of thin layer evaporator, crystallization pan and solar dryer. The improved process is simple and easily adaptable by the rural dwellers. This paper therefore reviews both the indigenous and the improved methods of salt production in Nigeria.

Keywords: *Natural brine, Salt production, Indigenous process, Thin layer evaporator, Crystallization, Physico-chemical characteristics.*

Introduction

Salt occurs naturally in many parts of the world, however, the nature of occurrence and processing affect the quality and yield of products and by-products. Salt is physiologically necessary for human life and it is an essential element in the diet of not only humans but of animals and plants. Its industrial and other uses are numerous.

In spite of the occurrences of natural brine in lakes, ponds and springs in Nigeria, the country still imports about 99 per cent of its annual salt requirement. It has an estimated consumption of 600, 000 metric tonnes per annum with per capita consumption varying between 2.2g and 6.3g a day and imports $2.3 billion worth of salt annually for domestic use and for its petrochemical industry (Malomo, 2011).

There is no cottage salt industry in the country, except the effort of the rural women in indigenous salt production. The salt produced is sold in the village markets within the salt producing communities, which means that only the people within these communities consume the salt.

This paper therefore reviews both the indigenous and the improved methods of salt production in Nigeria.

Nature of Salt Occurrences

Salt deposit is classified according to its formation. In nature, it occurs in solution or in a solid state. In solution, it is found in oceans, lakes and springs. It is generally believed that the bulk of the world salt supply is in solution in the oceans. The salt content of the oceans varies from 1 to 5 per cent. This variation which exists from place to place and by depth is due to the influence of in-flowing waters, variation in the evaporation and freezing rates. Seawater contains an average of 2.6 per cent (by weight) NaCl, or 26 million metric tons cubic kilometer (www.earthsci.org.mindep/salt/salt.htm).

Salt lakes are formed when water accumulates in a topographic depression. Waters flowing into the depression carry material in solution. Thus, it is possible to have lakes that vary greatly in composition. Sodium and calcium are the predominant cations in mineralized lake deposits.

Spring waters vary in salinity from pure water to those that are almost completely saturated and the chemical content ranges from practically nothing (zero) to as much as 254,000 ppm (Lefond and Jacoby, 1983). Salt may also be obtained in the form of highly concentrated brines from wells that tap underground stream flowing through salt deposits.

Solid salt is found in sedimentary deposits such as playa deposits, bedded salt deposits, diapirs and in domes. A playa is a flat-bottom depression found in interior desert basins and adjacent to coasts within arid and semiarid regions such as found in Borno state in Nigeria. It is periodically covered by water which slowly infiltrates into the ground water system or evaporates into the atmosphere causing the deposition of salt, sand and mud along the bottom and around the edges of the depression. The dissolved minerals in the lake result from the leaching of the rocks in the area surrounding the enclosed basin. Salt in the playas results from the solution of sodium chloride – bearing sediments.

Bedded salt deposits are true sedimentary rocks and as such are found in sedimentary sequences with shale, limestone, gypsum, anhydrite etc. Salt beds can be as much as 395.2m thick (Lefond and Jacoby, 1983). However, due to the manner in which they are formed, thick salt beds characteristically contain thin bands or lenses of anhydrite and/or clay, with a salt interval of 50.8mm to 101.6mm thick. In some areas as much as 60.8m of salt are known (Lefond and Jacoby, 1983).

The salt dome is a kind of salt formation, which is formed by the nearly vertical intrusion of salt upward through overlying rock deposits. Salt is very vulnerable to pressure, thus, as pressure is exerted on a salt bed due to static weight of overlying sediments or tectonic forces, the salt "flows" plastically, first bulging to form salt anticlines. If the pressure increases, the overlying rock is ruptured and salt is squeezed into, and through the overlying sediments to form salt domes or diapiric folds. In some instances, the salt reaches the surface or actually flows like a glacier onto the surface (Lefond and Jacoby, 1983).

The values of impurities found in various types of salt are shown in Table 8.1.

Table 8.1: Impurities in Natural Salt

Impurity	*Rock Salt (per cent)*	*Sea Salt (per cent)*	*Lake Salt (per cent)*	*Brine (per cent)*
$CaSO_4$	0.5–2	0.5–1	0.5–2	Saturated
$MgSO_4$	Traces	0.2–0.6	Traces	Traces
$MgCl_2$	–	0.3–1	Traces	–
$CaCl_2$	–	–	Traces	–
Na_2SO_4	–	–	Traces	–
KCl	–	–	Traces	–
NaBr	–	–	Traces	–
Insolubles	1–10	0.1–1	1–10	–

Source: Sedivy, (1996).

From the table, it can be seen that brine is saturated with calcium sulphate ($CaSO_4$) while other types of salt contain minimal amount. However, the magnesium, calcium, and potassium chloride contents are almost not detectable.

Brine Occurrences in Nigeria

Nigeria is blessed with natural brine in the form of lakes, saline springs, and ponds but there is no known rock salt deposit in the country. A number of saline springs, lakes and ponds occur in the Benue Trough within a relatively narrow belt which extends in a north-north-east to south-south-west (N.N.E-S.S.W) direction from Gombe to Ebonyi state, along the generally low lying and gently undulating plains of the Benue and Cross River drainage systems (Uma, 1998). Saline groundwater occurring mostly as springs, ponds, and in dug wells are found in three major geographical regions of the trough: northern, central and southern regions.

Brine Locations

The Nigerian brines are located in four main areas: Gombe, Shendam, Abakaliki and Ikom-Calabar where they have given rise to small local salt production centres. The location of the brine occurrences is presented in Table 8.2.

Table 8.2: Brine Locations in Nigeria

Area		*Locations*
Shendam		Abuni, Akiri, Akwana, Arufu, Awe, Azara, Kanje, Keana, Langrel River, Lankaku, Ayero, Ribi, Imon, Odzi, Akyanapole, Okumaka, Adaka
Abakaliki	Ogoja zone	Abatchor(W.Yala), Echimoga, Gabu, Ijegu, Igbekureku, Itegaokudu, Okircha, Okpame (W.Yala), Okpodo,Okpoma
	Ameri zone	Abakaliki, Ameka, Ishiagu, Ameri, Enyibichiri, Ikwo, Iyi-Oku River, Nkaliki, Idembia Ezza, Obegu, Ike
	Afikpo zone	Achara, Ezeukwu, Okayi, Okposi, Uburu
Ikom – Calabar		Abachor, Akparabong, Idigbo, Ijegu, Ibene, Oguwaru, Okpinyia, Okircha, Okpame, Okpoma, Igbekuleku, Olachor, Ore, Aduoji, Opupu, Ohoda, Ojeje, Abanerok, Abi (near Etomi), Danare, Ikom, Okuni, Odukpani
Gombe		Ayabe, Bumanda, Gyakan, Gujba, Jebero, Jebieb, Jendo, Langa, Mutum Daya, Pero Hills, Pindiga, Takulma, Todi, Gumsi, Nafada, Tumu, Zanga.

Source: Beltano and Bokarski, (1971).

Beside the four main areas considered above, there are other scattered occurrences of brine. They are found at Gumel (Kano state), Bale (Sokoto state), Moi-Igbo and Oku (Benue state), Bunza and Kamba (Kebbi state).

Characteristics of the Brines

The brine springs of Awe, Azara, Kanje and Keana in Nasarawa state were investigated by the Geological Survey of Nigeria (now Nigerian Geological Survey Agency, NGSA) in 1942 as shown in Table 8.3.

Table 8.3: Characteristics of Azara, Kanje, Keana and Awe Brines

Locality	*Temperature*	*Salt (%)*	*NaCl (%)*	*Remarks*
Awe spring	92º F	0.7	Predominant	Traces of magnesium and bicarbonates
Awe borehole	82º F	0.8	Predominant	Traces of magnesium and bicarbonate
Azara spring	82º F	2.0	Over 90	Traces of boron and iodine
Kanje	82º F	0.94	91	–
Keana	95º F	Very dilute	Predominant	Seepage

Source: Beltano and Bokarski, (1971).

The Table shows that Keana spring, Awe spring and the groundwater obtained from borehole were predominantly NaCl with traces of magnesium and bicarbonates. According to Tattam (1942), the brine samples collected from Okpoma and Igbekureku (Cross River state) show 74 to 76 per cent NaCl, the remaining percentage being composed principally of Ca, Mg and K salts. Tattam also gave the value of 81 per cent NaCl for brine in Ameri (Ebonyi state) which was obtained after a month of continuous pumping during dewatering of the Ameri lead-zinc mine.

Uma (1998) carried out an investigation on the chemical constituents of major Nigerian brines and the result is presented in Table 8.4.

Two water types are recognized, those with electrical conductivities (EC) less than 1000 µs cm are generally rich in bicarbonates, alkaline earth metals, calcium and magnesium. This water type was taken to represent the normal groundwater type in the Benue Trough (unaffected by contact with the up-fluxing brines). The second group with EC over 1000 µs cm is rich in alkalis (sodium and potassium) and constitute over 80 per cent of the dissolved constituents. This group may be taken to represent brines with various degrees of dilution (arising from contact/mixing with the normal groundwater type).

Indigenous Salt Production from Brine

Salt production has been an indigenous technology in Nigeria, especially, among women who are making salt from brine in lakes, springs, ponds and saline earth. The production of salt from brine has been in existence for over 200 years. It is entirely done by women through crude and laborious method of evaporation in open fire using firewood. Several occurrences of brine have given rise to small local salt production centres in different localities within the country.

Salt Production Processes

The production of salt from brine spring, lakes and ponds is a practice that has been in existence for over a century and differs slightly from place to place. The process is carried out only during the dry season. In the salt producing areas, salt processing starts in November at cessation of the rains and continues until the onset of the rainy season in April or early May of the following year. The operation is done entirely by women and the salt production processes are as follows:

Collection of Brine from Lakes, Ponds or Spring Pools

The women assemble in the morning each day and use containers to carry accumulated brine from the spring pool, lakes or ponds (Figure 8.1).

Concentration of the Brine

The collected brine is sprinkled on thin layers of mud already prepared on a hard mud floor divided into compartments (Figure 8.2) until the soil becomes soaked with considerable amount of brine. In the afternoon, the women assemble again to gather the soil which must have been dried up by the heat of the sun. The salt at this stage is an accumulation of both soil and salt crystals which is then collected by scraping the soil surface to which salt adheres. The rate at which the salts accumulate

Table 8.4: Major Chemical Constituents of Nigerian Brines

Sample Location	*Date*	*EC(μs cm)*	*Ca^{+2}*	*Mg^{+2}*	*Na^{+}*	*K^{+}*	*SO_4^{-2}*	*HCO_3^{-2}*	*Cl^{-}*	*Source Type*
Enyigba	18/07/90	12160	6.03	1.78	101.0	2.01	–	9.4	100.00	Shaft
Okposi	05/05/91	49600	39.9	9.90	923.5	16.4	–	3.9	902.7	Pond
Okposi	05/05/91	680	3.9	1.0	2.3	0.2	0.4	6.2	0.7	Shallow borehole
Uburu	05/05/91	69700	49.4	14.8	1000.5	14.1	–	4.8	987.4	Spring
Ishiagu	20/03/94	10250	12.7	1.8	78.8	2.5	–	3.9	79.0	Mining pit
Ishiagu	20/03/94	257	1.2	1.3	0.1	0.1	–	0.4	0.0	Abandoned mine
Abatchor	04/03/94	108400	202.6	26.8	1092.0	16.1	–	1.4	1445.0	Dug hole
Okpoma	04/03/94	390	3.8	1.5	1.2	0.0	–	4.3	0.0	Shallow borehole
Ijegbu Yalla	04/03/94	99300	143.9	18.5	1023.0	11.2	–	3.5	1021.9	Natural hole
Kanje	01/05/91	11500	7.5	4.0	108.8	3.1	–	2.1	115.7	Spring
Kanje	01/05/91	21500	13.5	4.1	213.2	4.6	–	6.2	228.5	Spring
Awe	30/07/90	15000	6.6	2.8	132.5	1.9	–	10.1	135.1	Spring
Awe	30/07/90	14990	6.6	2.8	131.9	1.9	–	10.0	134.0	Warm spring
Ribi	02/05/91	17010	14.0	3.9	165.3	3.8	–	8.8	172.1	Pond
Ribi	02/05/91	1500	1.6	0.6	13.9	0.2	0.9	11.9	3.7	Dug well
Azara	02/05/91	28200	22.0	6.8	282.8	5.9	–	10.0	310.3	Dug hole
Kaseyo	03/05/91	509	8.5	0.9	0.6	0.3	6.9	3.0	0.1	Spring
Tse Tavaan	03/05/91	6700	7.5	4.0	56.6	0.8	2.5	1.6	64.9	Spring
Moi Igbor	27/02/93	1800	1.38	0.5	14.3	0.5	–	2.0	11.9	Spring
Moi Igbor	27/02/93	1100	2.9	2.8	6.5	0.1	2.6	8.4	0.1	Dug well
Moi Igbor	07/03/94	5190	1.5	0.5	47.2	0.6	–	11.2	29.6	Seepage
Ocean Weter at PH	1980	–	8.1	80.6	443.7	11.0	53.8		510.0	Atlantic ocean
Ocean Water at Opoko	1980	–	8.4	81.0	487.2	10.7	52.7		506.2	Atlantic ocean

Source: Uma, (1998).

Figure 8.1: Natural Brine Pond.

Figure 8.2: Hard Mud Floor Divided into Compartments.

on the soil depends on the salinity or concentration of the brine while the formation of the salt crystals on the surface layer of the soil depends on the intensity of the sun. The soil is then collected and prepared for the filtration process.

Leaching of Salt from Saline Earth

Large perforated earthen pots are partly filled with the mud/sand and brine from the spring pool is added to re-dissolve the salt. A highly concentrated brine drips continuously from the hole at the bottom of the pot into a receptacle (Figure 8.3).

Boiling/Crystallization of the Salt/Sun Drying

The concentrated brine is later carried to nearby mud-wall, thatch shelters for boiling. The brine is poured into cooking pots and boiled using firewood. The boiling continues in the constructed huts (Figure 8.4) until the water evaporates leaving crystallized salt.

The final drying of the salt to reduce its moisture content is achieved through sun drying. The harder part of the salt at the bottom of the pot is used mainly for animal leak. The crude salt (Figure 8.5) is packed and preserved to prevent dissolution by atmospheric moisture (Figure 8.6). The salt is ready for consumption and is packaged in woven mats for the market (Figures 8.7a-b).

The entire processes from concentrating the brine through evaporation to packaging are crude, laborious and time-consuming and the result is an impure mixture of insoluble solids and various soluble salts in which sodium chloride (NaCl) is the dominant constituent. The quantity of salt produced is not commensurate with

Figure 8.3: Leaching of Salt from Saline Soil.

Figure 8.4: Huts where Brine is Boiled Salt.

Figure 8.5: Sample of Crude Salt.

Figure 8.6: Packaged Salt.

Figure 8.7a: Sample of Crude Salt.

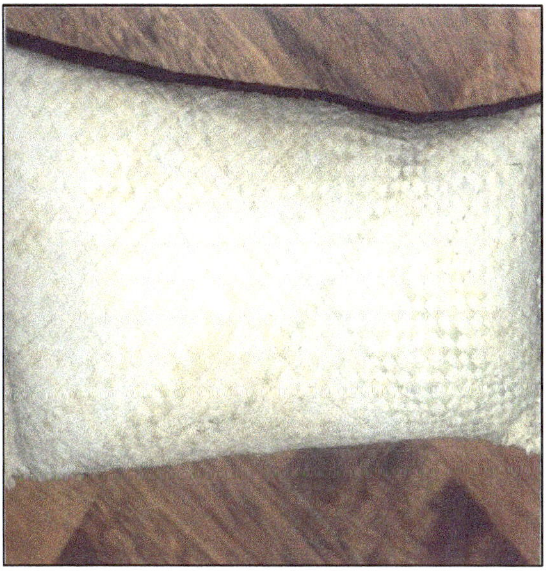

Figure 8.7b: Crude Salt Packaged in Woven Mat.

the time spent and the effort put into the production process as the women produce about 12 – 15kg of salt in 28 days.

Quality of the Salt

The quality of the salt obtained from these brines is fairly good. Sodium chloride is by far the most important constituent and the amounts of impurity in the form of calcium, magnesium and potassium or the sulphate and carbonate salt are usually small and insufficient to impair the value of the salt. However, analyses made at the laboratory of Imperial Institute, London in 1942 on samples of salt prepared by the salt producers using the indigenous method seem to indicate that these are not the representative of the salts of the brines.

It is presumed that owing to filtration process, the locally made salt may contain unduly high percentage of certain constituents. For example, salt produced by the local salt producers at Uburu, Ebonyi state, which was analysed at the Imperial Institute, indicated about 7 per cent potassium chloride (KCl) whereas the analysis made on salt prepared in the laboratory from the natural brine contained only 1.16 per cent (Tattam, 1942).

Likewise, analysis for salt samples produced in different parts of Awe, Nasarawa state showed a range of potassium chloride content from 2.8 - 8.2 per cent whereas a sample of well powdered and mixed salt obtained by complete evaporation of the brine in a laboratory gave only 1.6 per cent (Tattam, 1942).

It is possible that the leaching process used by local salt producers could be associated with the increase in potassium chloride content and high level of impurity in the salt.

Brine and Saline Earth Used in the Leaching Process

The result of the chemical analyses of brine and saline earth used during the leaching process is presented on Table 8.2.

The table shows that the natural brines from various locations contain some amounts of potassium, calcium and magnesium salts. However, the values obtained from the saline earth samples are highly significant when compared with the natural brine. It is important enough to have an effect on the resultant concentrate.

Also, metals such as lead (Pb) and iron (Fe) which were not detected in the natural brine had significant values in the saline earth samples. This could probably be the source of the impurities in the crude salt.

Challenges of Indigenous Salt Production in Nigeria

The major problems identified in course of salt production in Nigeria, include the following:

- Access to the brine is difficult in some places due to the fact that it is obtained from pits/ponds that are susceptible to slope failure and flooding.
- Availability of the brine is seasonal. Concentrated brine can only be obtained during the dry season (January – April) when the brine in ponds has been concentrated by sunlight.
- Local salt production is laborious and risky, especially, for aged women who are the major processors. It is characterised by wasteful and inefficient practices.
- The processes involved are energy and time consuming. It involves vigorous boiling of large volume of brine for a long time usually 2-3 days to produce a batch of salt.
- The crude method used in concentrating the brine introduces insoluble solids and other impurities into the brine.
- The local salt producers have limited financial means and lack access to technical assistance. As a result, the salt produced is of poor quality and is visibly contaminated.

Table 8.2: Chemical Characteristics of Brine and Saline Earth from Nasarawa and Kebbi States

Components	*Brine*					*Saline Earth*				
	Azara	*Awe*	*Bunza*	*Kamba*	*Keana*	*Azara*	*Awe*	*Bunza*	*Kamba*	*Keana*
Na (per cent)	0.52	0.24	5.04	0.15	0.17	1.39	1.28	2.5	1.51	1.35
K (ppm)	226.11	83.66	1920.2	98.05	96.82	903.3	959.0	1329.49	691.74	1998.57
Ca (ppm)	145.88	37.38	243.0	150.05	50.66	720.47	10929.2	5815.8	1015.28	8574.0
Mg (ppm)	58.48	10.93	3087.2	74.47	3.23	1434.6	2114.8	2719.0	1061.91	1692.0
Fe (ppm)	ND	0.27	3.62	0.7	0.36	22710.8	9853.4	18721.4	5506.01	28173.6
Zn (ppm)	ND	ND	0.48	ND	ND	23.84	14.43	18.74	14.33	24.03
Cu (ppm)	0.01	0.02	0.07	0.01	0.02	5.54	5.11	5.8	3.24	4.7
Pb (ppm)	ND	0.15	ND	ND	ND	487.17	44.0	76.57	694.91	52.27
Cd (ppm)	ND	ND	ND	ND	ND	ND	ND	ND	0.38	ND
CO^- (ppm)	ND	1.14	3.45	2.3	1.15	ND	ND	ND	ND	ND
HCO_3 (ppm)	5.84	18.7	67.78	14.02	9.55	ND	ND	ND	5.84	ND
Cl^- (per cent)	0.85	0.4	6.93	0.24	0.3	2.47	2.34	4.25	2.35	2.17
$NaCl^-$ (per cent)	1.32	0.61	11.44	0.38	0.43	3.53	3.25	6.35	3.84	3.43
S.G	1.01	1.0	1.14	1.0	1.0	2.68	2.65	2.59	2.6	2.62

- ☆ The profit margin from the production is very minimal, hence the process is uneconomical. This explains why younger women are not involved in salt processing and it may gradually go into extinction.

Efforts of the Raw Materials Research and Development Council (RMRDC) in Improving the Indigenous Salt Production Technology

In 2003, RMRDC collaborated with the Federal Polytechnic, Nasarawa state, Nigeria to upgrade the indigenous salt production technology through the fabrication of simple equipment which would mimic the indigenous process used by the women.

The simple machinery to carry out the upgraded salt production techniques for a small-scale cottage production consists of the following equipment:

- ☆ Thin Layer Evaporator (TLE)
- ☆ Crystallization Pan
- ☆ Solar Dryer

The flow diagram for the upgraded salt process technology is given below:

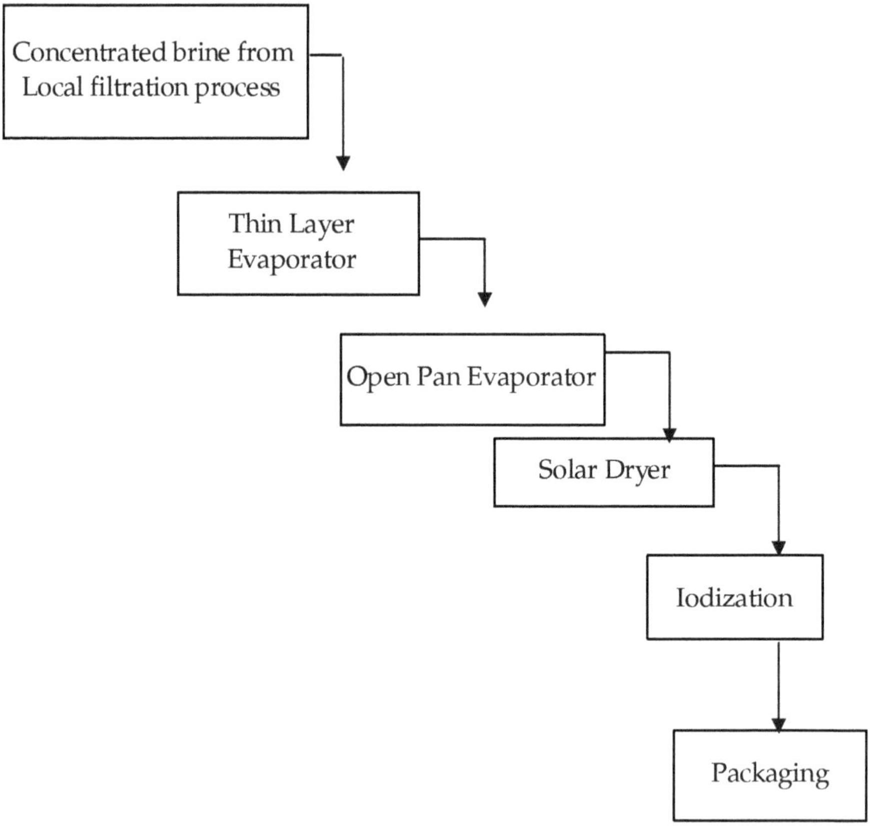

Figure 8.8: Flow Diagram of Upgraded Salt Production Processes.

The Process Technology

The concentrated brine from the local leaching process is poured into a storage tank. From the tank, it flows through a pipe into the thin layer evaporator (TLE), where the brine is heated. The heat source could be firewood, coal briquette or rice husk briquette enclosed in a heating chamber.

There is a retention period or residence time of about 12 – 18 minutes before the brine flows to the end of the TLE. At this point, a sieve is fixed at the flowing end of the evaporator to remove the insoluble solids before the brine is collected in a plastic bucket.

The brine is recycled for about 3 – 4 times on the TLE for further concentration until salt crystals begin to form. As soon as the brine becomes liquor with enough salt crystals, it is allowed to flow into the crystallization pan through a conveyor. The salt crystals formed on the thin layer evaporator during the process of boiling are shown in Figure 8.9.

Fresh brine is poured into the storage tank and the whole process is repeated. Meanwhile, at the discharge end, the wet salt is left in the open pan for further crystallization until sufficient crystals have been formed. The liquor decanted from the salt crystals is collected and returned to the thin layer evaporator while the salt crystals are distributed in trays and stacked in the compartments within the solar dryer for final drying to reduce the moisture content.

Figure 8.9: Salt Crystals Formed on the Thin Layer Evaporator during Production.

The equipment enhances the quality and quantity of the salt but the crude salt produced at this stage, requires refining to remove other impurities and dirt. The equipment was designed to have the capacity to produce 60kg wet salt in 8 hours.

Table 8.5 shows the capacity of the various units of the equipment while (Figure 8.10) shows the installed plant at a site in Ebonyi state.

Table 8.5: Capacity of Each Unit of the Equipment

1.	Thin Layer Evaporator (TLE)	7.5kg/hr of wet salt
2.	Solar Dryer	15kg of dry salt/2 hrs production

Figure 8.10: Installed Salt Production Plant.

Way Forward

The Raw Materials Research and Development Council in collaboration with the Nasarawa State Government held a stakeholders' forum in June 2012, to intimate the stakeholders on its readiness to deploy the upgraded salt equipment in empowering the existing salt processing clusters in the state. Also, due to the importance of the project to the Nigerian economy, the possibility of establishing a salt refining and iodization plant in the state was discussed. The Nasarawa state government has provided two units of the salt equipment installed at Awe and Keana with access roads and building structures. Other facilities are to be provided for the salt processing communities.

Schematic Diagram of the Salt Cluster Networking

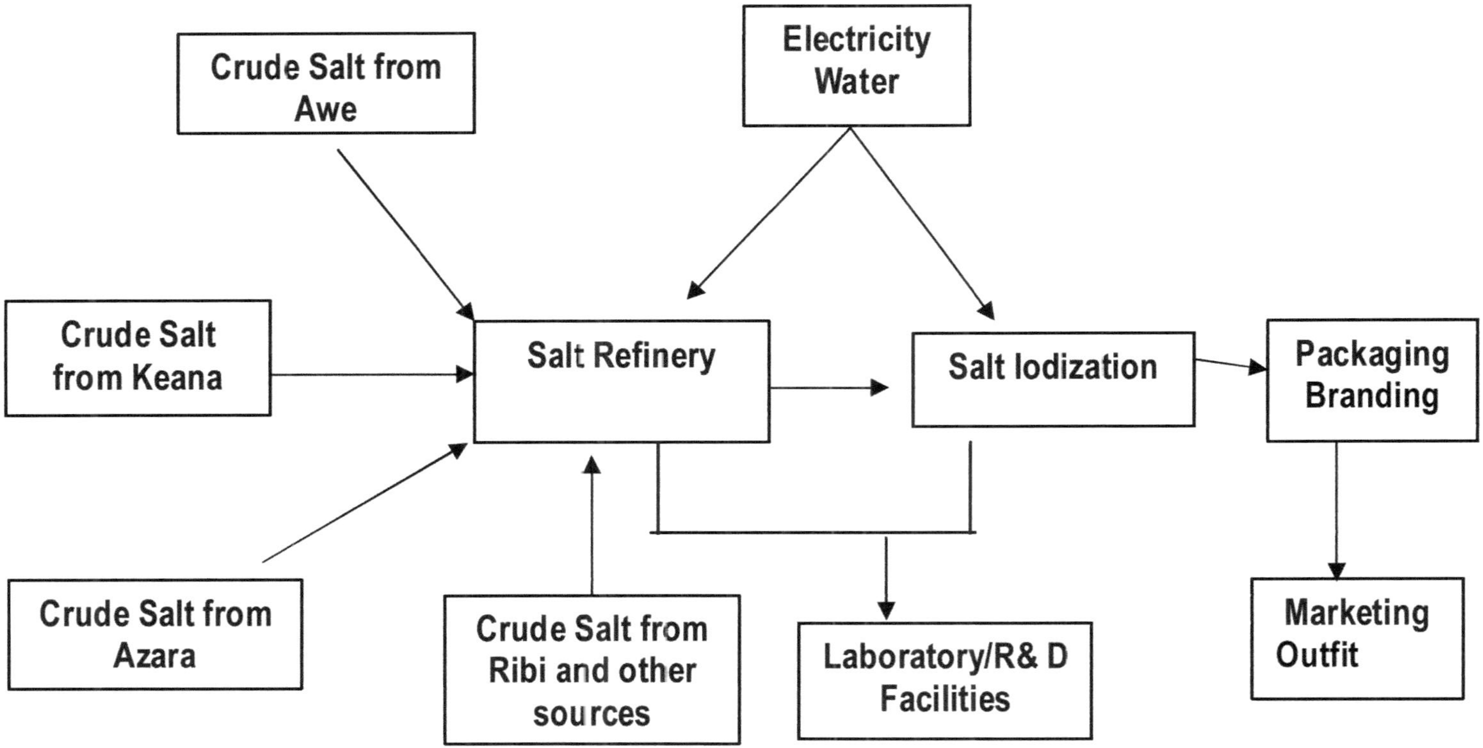

Figure 8.11: Salt Cluster Networking in Nasarawa State.

Conclusion

Salt production in rural communities in Nigeria is still being carried out by numerous indigenous producers without refining and iodization and the rural women involved in the processing are currently facing serious challenges. In other to alleviate the current challenges faced by these women, there is need for investors to invest in replication of the upgraded salt equipment as well as establish salt refining and iodization plants in Nigeria.

Acknowledgement

We wish to acknowledge the management of the Raw Materials Research and Development Council for sponsoring this project. Our sincere appreciation also goes to Prof. Bugaje, former Rector of the Federal Polytechnic, Nasarawa and entire Engineering Team of the Chemical Engineering Department for their efforts toward improving the equipment design

References

1. Beltano F and Bojarski R (1971). The Possibility of a Salt Industry in Nigeria. Geological Survey of Nigeria Report N0. 1492, pp. 2 – 14.
2. Lefond SI and Jacoby CH (1983). Salt. Industrial Minerals and Rocks (Non-metallics Other than Fuels). Fifth Edition, Vol. 2, pp. 1119-1149. New York, USA.
3. Malomo S (2011). Mineral Production and the Challenges of Value Addition: An Overview. Paper Presented at Nigerian Mining and Geosciences Society/ Nigerian Geological Survey Agency Discourse, 2011, pp. 7.
4. Sedivy VM (1996). Purification of Salt for Chemical and Human Consumption. Industrial Minerals Yearbook- 1996. pp. 1-20. KREBS SWISS, Zurich, Switzerland.
5. Tattam CM (1942). Preliminary Report upon the Salt Industries: Eastern Nigeria. Geological Survey of Nigeria, (GSN) Report No. 778, pp. 1–8.
6. Uma KO (1998). The brine fields of the Benue Trough, Nigeria: A comparative study of geomorphic, tectonic and hydro chemical properties. *Journal of African Earth Sciences*. 26 (2), pp. 261 – 275.
7. Salt – The Salt Institute, USA. http://www.earthsci.org/mindep/salt/salt.htm (accessed 25 Oct. 2004).

Chapter 9

Challenges of Mining, Processing and Beneficiation of Columbite in Nigeria

Elizabeth O. Wuyep[1], Moses O. Omojola[1], Peter A. Onwualu[1] and Kehinde Ogunmola[2]

[1]Raw Materials Research and Development Council, Abuja, 900001, Nigeria. PMB 232, Garki, Abuja, Federal Capital Territory, Nigeria
[2]National Centre for Remote Sensing, Jos Consultancy Services, Plateau, 900001, Nigeria
E-mail: ewuyep@yahoo.com, omojolamoses@hotmail.com, ponwualu@yahoo.com and kennybonzo@yahoo.com

ABSTRACT

The Pan-Africa (600 ± 150Ma) basement of Nigeria is known to host rare-metal pegmatites containing economic concentrations of columbite-tantalite ore minerals. These pegmatites are found in the NE-SW trending belt stretching from Jos Plateau through Isanlu-Egbe in Central Nigeria to Ijero – Aramoko – Ilesha area of southwest Nigeria. The columbite if properly exploited can lead to national economic growth. Unfortunately, little information is available on the deposits and the present mining is still at artisanal level due to lack of modern and innovative research, lack of advanced skills and modern equipment. This paper focuses on the above challenges and how these challenges have affected the product yield leading to massive loss of revenue as a result of improper processing techniques. Also, of much concern is the fact that not much is known of the total current reserve, the geographical distribution, the mode of geologic occurrence, and the quality and physicochemical properties of columbite in Nigeria.

***Keywords**: Columbite, Mineral, Environmental challenges, Mining, Processing, Beneficiation.*

Introduction

Nigeria occupies an area of 923,773 square kilometer which falls within latitude $10^{\circ} 0^{2} 00^{3}$ N and longitude 8° 0' 003 E. It is located in West Africa and bounded to the north by Niger, to the east by Chad and Cameroon, and to the west by Benin (Figure 9.1). The total population size of Nigeria is over 141,000,000 (National Population Commission, 2007) with a population density of 173.94.

Nigeria is endowed with many rich mineral deposits one of which is columbite that if properly exploited can lead to wealth creation, hence, contribute immensely to improving the overall quality of life as well as value creating asset. Unfortunately, little is known about the deposits and mining is still at the artisanal level, this is attributed to the fact that: much attention is focused on oil and gas sector; there is lack of advanced skills and modern equipment and there is also lack of modern and innovative research. Despite the efforts made by stakeholders, little progress has been made in terms of policy reforms and recreation of the enabling environment for these columbite deposits to be positively exploited for industrial development.

The study area Jos and its environs (Kuru and Rayfield) lie in the Middle Belt of Nigeria. It is about 104Km from North to South, and 80km from east to West covering an area of 8,600km^2 (Ndace and Danladi 2012). Jos Plateau is situated between Latitudes 9° 302 N and 10° N and longitudes 8° 302 E and 9° E. Geologically, Jos falls within the Nigerian North Central Basement Complex which occupies about 50 per cent of the Country (Figures 9.2 and 9.3). Younger Granites and Sedimentary rocks also form the other components of the Nigerian geology. Of the three components of

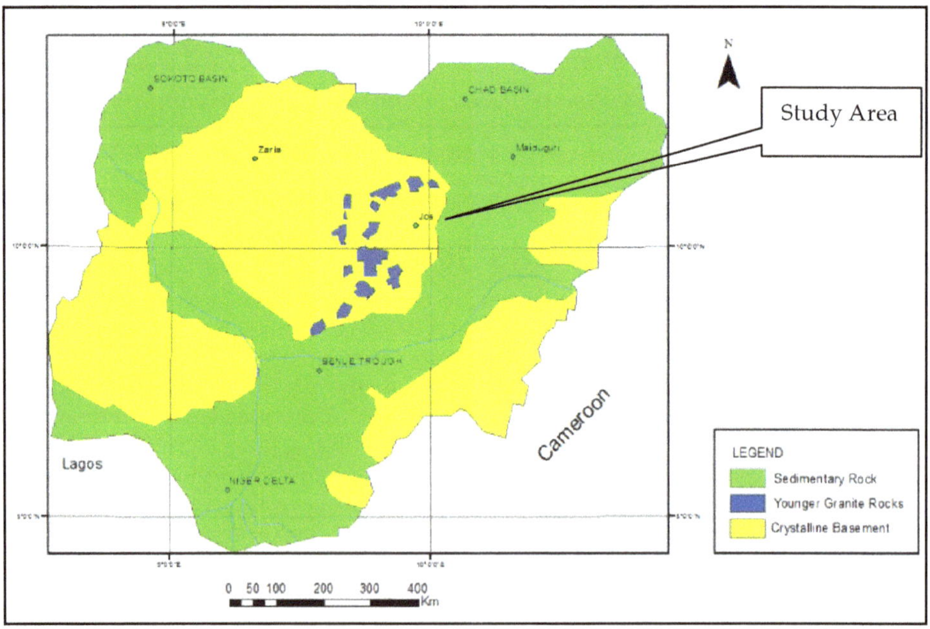

Figure 9.1: Simplified Geological Map of Nigeria Showing Study Area (after Kogbe, 1985).

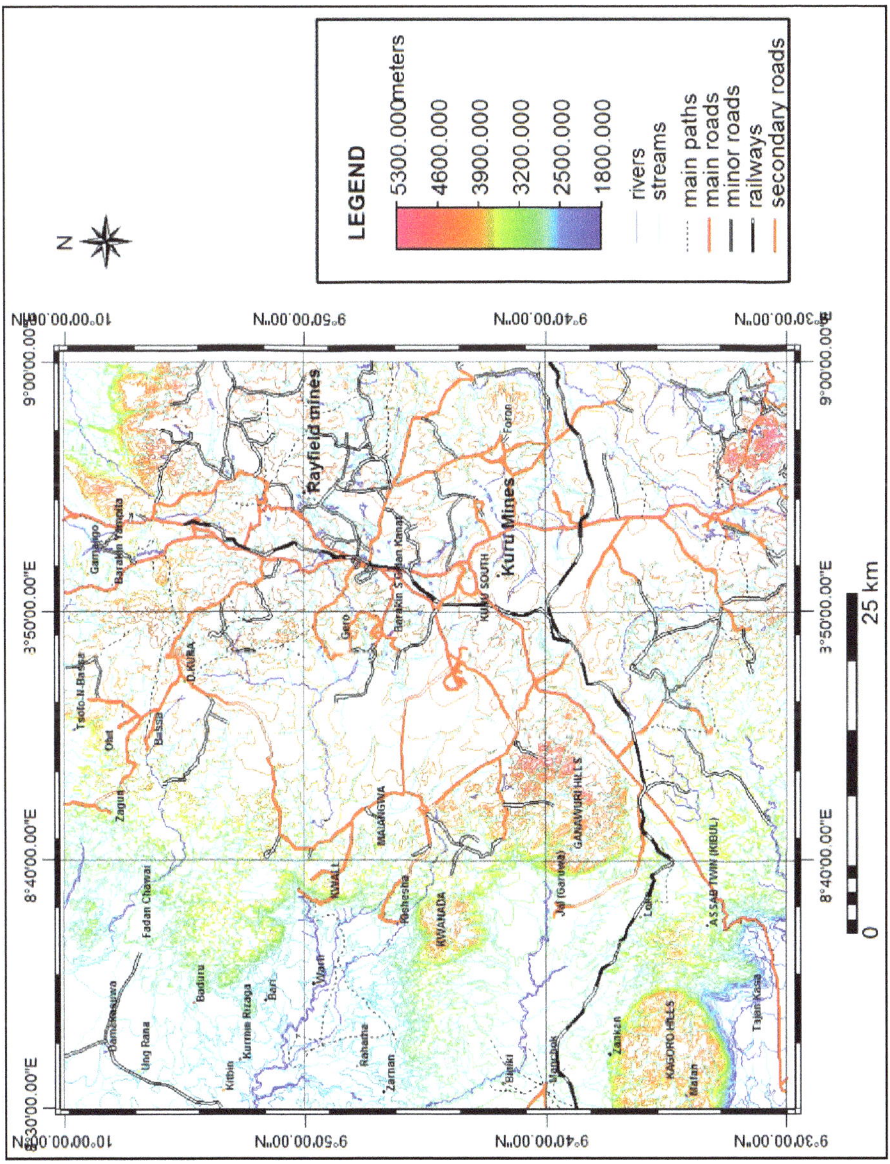

Figure 9.2: Topographic Map of the Study Area Showing the Location of the Rayfield and Kuru Mines.

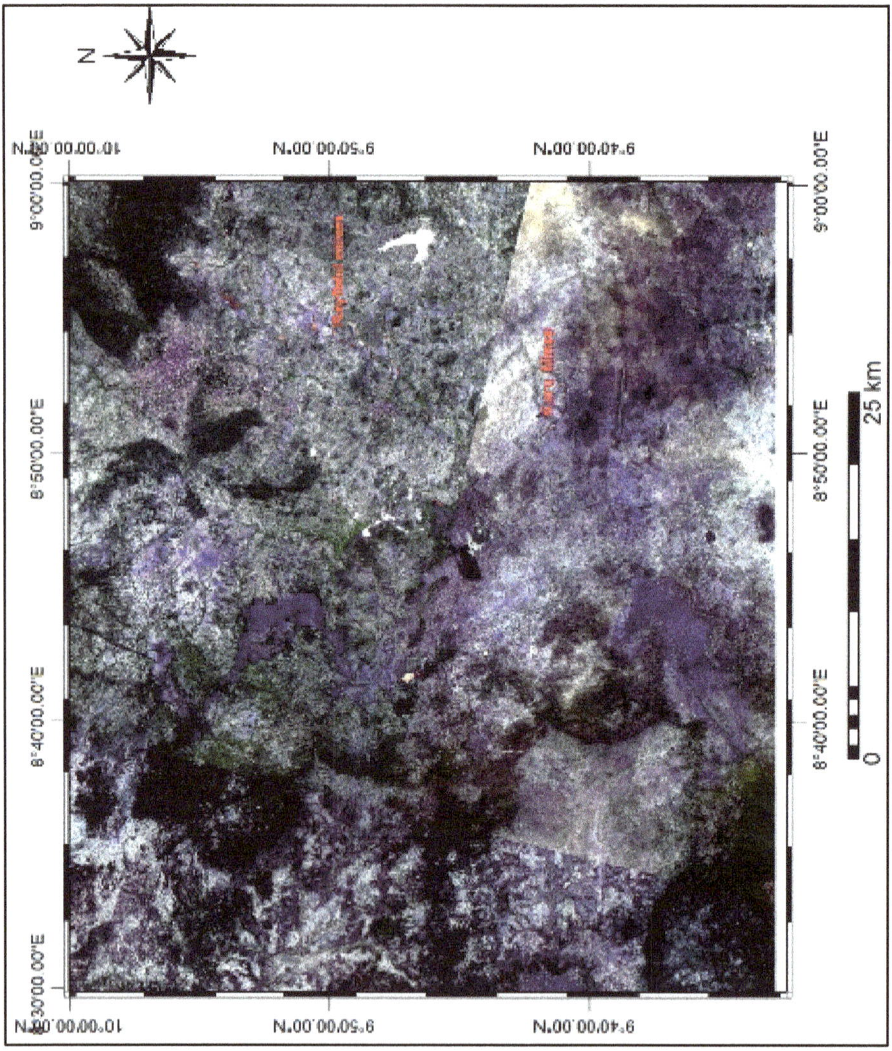

Figure 9.3: Spot 5 Image of the Study Area Showing the Location of the Rayfield and Kuru Mines.

the Nigerian geology, Plateau state is covered by the Younger Granites and Basement Complex rocks only. Both rocks show a general N – S, NW - SE and NE - SW structural trends, those of Basement Complex seem to lie parallel to the limbs of major folds.

Occurrence of Columbite in Nigeria

Columbite, a black mineral group is an ore of Niobium that is high in density with a metallic luster (Figure 9.4). It is a high density mineral and a niobate of Iron and Manganese [$(Fe, Mn)Nb_2O_6$]. The Pan – Africa (600 ± 150 Ma) basement of Nigeria is known to host rare-metal pegmatites containing economic concentrations of columbite-tantalite ore minerals. These pegmatites are found in the NE – SW trending belt stretching from Jos Plateau through Insanlu – Egbe in Central Nigeria to Ijero-Aramoko – Ilesha area of southwest Nigeria. Columbites are known to occur in Plateau, Kaduna, Bauchi, Kogi, Kwara and Nasarawa states. Columbites and tantalites occur together in granites, pegmatites and placer deposits. In Jos Plateau and environs, columbite is hosted in biotite-granites of the Younger Granites and occurs in association with cassiterite, monazite, zircon and ilmenite. Macleod *et al.* (1971) discussed the accessory minerals in biotite and riebeckite granites in the Younger Granites of Jos Plateau. In Nigeria, the Younger Granites are found mainly

Figure 9.4: Columbite from Jos Plateau.

on the Jos Plateau and some parts of Nasarawa state, they form a distinctive group of intrusive and volcanic rocks bounded by ring faults or dykes. Macleod *et al.* (1965) and Alkali and Yusuf (2010) have provided reviews of the geology of the Younger Granites. It is worthy of note that not much is known of the total current reserve, the geographical distribution, the mode of geologic occurrence, quantity and quality and physicochemical properties of columbite in Nigeria.

Mining

Mining of solid mineral in Nigeria started as far back as 1903 on the Jos Plateau with tin mining in the upper parts of Plateau and Lead-Zinc in the lower Plateau that includes some parts of Nasarawa state (Eckert *et al.*, 2001). Mining of tin has been largely responsible for profound changes in the Landscape, the social and economic structure of the Jos Plateau over the last 50 years. The mining region today can be as one of the several ecological sub-regions of the Jos Plateau (Ndace and Danladi 2012). Ajaegbu (1985) stated that the delimitation of the mining region is based on the area extent of past and present mining activities, the distribution of identifiable mining landscape features and mining settlement. However, according to Ndace and Danladi (2012) when the bulk of Nigerian tin was found by the year 1902 in Jos Plateau, the mining industry provided employment opportunities and most importantly the mineral exploitation provided a good source of revenue. Information was first received about the presence of tin on the Jos Plateau then known as Bauchi Plateau in 1884, but it was not until the Mineral Survey of Northern Nigeria (1904 – 1909) proven the occurrence over wide area of the Plateau (Hodder, 1959). Mining is still at artisanal level and it is done manually with the use of shovel, hoe and digger (crude method of mining) with no formal training of the miners. It is in view of these that this present study is carried out. In developed countries, before mining of any mineral is embarked upon, study is carried out to determine the above parameters and mining is done mechanically. This helps them to know the end from the beginning. In the past, columbite was seen as an accessory mineral and produced as concentrates from tin slag, but today the story is different as it is being mined as ore. Mining is done by surface mining (Open pit mining) which involves excavation with the use of digger, shovel (Plate 9.1) and also transported to the surface manually using shovel (Plate 9.2). The ore is separated from the tailings by washing, a process called gravity separation (Plate 9.3) until the ore get concentrated (Figure 9.4). This local mining requires about 6 – 10 people in the pit at once (Plate 9.5) and these workers and their counterparts in Democratic Republic of Congo (www.sda-uk.org/columbitetanta.htm) are paid $281 and $200 per month respectively, while their counterparts in South Africa earn $1,385, USA, $1,508, Japan, $1,260 (www.macon.com/southafricaminners), (www.favorimineral.com) and (www.minigreview.com). Despite the crude method of mining, about 30 – 50 bags of 50 kilogram columbite ore are obtained every week.

Processing/Beneficiation

Mining of mineral deposit produces raw minerals which need to be processed further by a specified method depending on the materials and waste product(s)

Plate 9.1: Open Pit Mining of Columbite at Kuru – Jos.

Plate 9.2: Workers Bringing the Ore to the Surface Manually for Washing Using Shovel.

Plate 9.3: Gravity Separation of Ore from the Gangue through Washing.

Plate 9.4: Gravity Concentration of Ore.

Plate 9.5: Showing the Number of Miners in the Pit at a Time.

associated with it. It is always necessary to have an intimate knowledge of the mineralogical composition of the ore for efficient processing to be carried out.

Generally, processing of columbite involves pre-concentrate and concentrate clean-up. The concentration processes which may be wet or dry gravity, magnetic or electrostatic methods produce concentrates that contains up to 70 per cent combined pentoxide to meet extraction requirements. However, the workers in the study area employ magnetic and dry gravity technique which is the technique employed universally. Magnetic separator is used to separate columbite from the gangue which include sand, tin, zircon, ilmenite and monazite (Plate 9.7), while Air float is used to separate tin, zircon and monazite from sand. Columbite is further concentrated using the Air float (Plate 9.8). With these equipments they are able to produce 2 tonnes of columbite per day and the staff strength is just four (4) who earn N700 per hour ($4.40), as a result of this, the demand is always higher than the supply.

To upgrade the metal concentration, the ores are beneficiated for further use in industries. There used to be a smelting company in Jos that was used to beneficiate tin in the past, but due to lack of maintenance it is not functioning any more. As a result, columbite is sold and exported after processing.

Market

About 87.60 per cent of the total worldwide requirements (50.6 Mlb of contained Nb) for niobium was marketed in 1999 in the form of ferroniobium for the high strength low alloy (HSLA) steel and stainless steel industries (Eckert *et al.*, 2010).

Due to low level of education, the miners are being exploited by informed people (Vendors) who buy the columbite from them at the rate of N4, 000 ($25) per kilogram (kg).These staff never received proper training, but have just General Certificate of Education/Senior Secondary Certificate of Education (GCE/SSCE). This method of processing is able to give a total yield of 36.5kg of pure columbite from 50kg of ore.

Challenges

The challenges associated with mining and processing/beneficiation in Nigeria include; Effect of mining on the environment, use of improper equipment, lack of skilled man power, low wages and Lack of formal market. These challenges will be discussed in detail the next section.

Materials and Methods

Materials

The data sets for this study consist of a topographic map of the Jos-Bukuru Younger Granites Complex and it's environment on sheet 168 Naraguta Southeast, global positioning system (GPS), Computer hard and software, Toyota Corolla, Inductively Coupled Plasma Optical Emission Spectrometer (ICP-OES), Spot Image, water sampling bottle, soil sampling bags, thermometer, Nitric acid and a PH meter. The main software used in the analyses of data is Integrated Land and Watershed Information System [(ILWIS 3.3) (2005)].

Plate 9.6: Magnetic Separator (Rapid) that Separate Columbite from Iron and other Metals.

Plate 9.7: Another View of the Magnetic Separator (A resource person standing to scale).

Plate 9.8a: Demonstration of Magnetic Separator as Gangue Drop Down the Conveyor Belt.

Plate 9.8b: Collection of Columbite from Magnetic Separator.

Plate 9.9a: Airfloat that Separate Columbite from Zircon, Tin and Sand.

Plate 9.9b: Airfloat Machine (A resource person standing to scale).

Plate 9.10: A Mine Pond in Rayfield on the Jos Plateau.

Sampling and Sample preparation

Eight samples were collected in all from three different sample media, namely, water, soil and ore from Kuru and Rayfield. The soil was randomly collected from designated localities at a depth of 10 – 15 cm and stored in labeled polythene bags.

The water samples were collected from the mining ponds in clean labeled plastic containers and immediately acidified with analar grade concentrated nitric acid (HNO_3^{2-}) to prevent loss of volatile metals and to inhibit bacterial and fungal growth. Determination of temperature and PH of water samples took place at the time of collection.

The soil samples were oven dried, crushed and pulverized to less than 90 micro-mesh using an agate mortar. Precisely 100 mg of the sediment powder was weighed into a teflon crucible using a triple-beam balance. Water samples already in solution form were filtered using No.42 ashless filter paper prior to analysis on ICP-OES (Inductively Coupled Plasma Optical Emission Spectrometer).

Sample Decomposition Procedure

100mg of the powdered sample was weighed into a Teflon crucible and dissolved in aqua regia ($HCl + HNO_3 + HClO_4$ (3:2.1)) after 6 hours of heating the solution to dryness on a hot-plate. After the crucible containing the sample was allowed to cool, 2mls of 2M HCl was added and topped with de-ionized water to about ¾ full and return to the hotplate to warm for not more than 15 minutes. The crucible + sample solution was removed from the hot plate and allowed to finally cool. The content was diluted to 100mls and filtered into a flat bottom flask ready to be run on the ICP-OES.

Methodology

Using direct observation, physical parameters such as colour and turbidity were determined. The heavy metals in the soil and water samples As, Be, Co, Cr, Cu, Fe, Mn, Ni, Pb and Zn were determined using Induced Coupled Plasma (ICP-OES). ILWIS was applied in digitizing, correction, classification, extracting and processing the image, and proper positioning of prominent features using UTM system.

Both the soil and water samples were analyzed by Inductively Coupled Plasma Optical Emission Spectrometry (ICP-OES). The instrument was calibrated prior to the introduction of sample by measuring in-house standards and blank solutions. The following major and trace elements were analyzed; As, Be, Co, Cr, Cu, Fe, Mn, Ni, Zn and Pb.

The quality of the analysis was controlled through the analysis of samples of known compositions along with the unknowns. The values obtained by the instrument from the reageant blank were automatically subtracted from the raw data before print out. This made the data free from whatever impurities inherited from the reagents including distilled and de-ionized water.

This work was carried out in a GIS environment, namely ILWIS. In general, remote sensing is the practice of deriving information about the earth's land and water surfaces using images acquired from an overhead perspective, using electromagnetic radiation in one or more regions of electromagnetic spectrum, reflected or emitted from the earth's surface. Spot 5 image covering the area of study was processed in the laboratory of the National Centre for Remote Sensing, Jos. A geometric correction was carried out on the image by changing the projection to UTM (Universal Transverse Mercator), the spheroid to Clarke 1880, Datum to Minna, Nigeria and Zone to 32. Several image processing techniques such as image enhancements and radiometric enhancements were carried out on the image which eventually enhanced the features on it. The processed image of the study area was displayed on the screen and properly studied by zooming in and out to clearly define structures/lineaments that are visible on the image. The mine sites were identified on the image as white spots which signify areas that are bare surfaces. An up-to-date topographic and road map was not available and had to be generated using data extracted from the digital elevation model (DEM) and satellite image of the study area. The contours were generated from SRTM (Short Radar Terrain Model) data at 30m interval while the road network and drainage pattern were extracted from the Spot 5 image.

Results and Discussion

Sample	*pH*	*As 198.69*	*Be 313.10*	*Co 228.61*	*Cr 269.716*	*Cu 327.39*	*Fe 238.20*	*Mn 257.61*	*Ni 231.604*	*Pb 220.35*	*Zn 206.61*
L1X1	6.6	0 mg/L	ND	0.001 mg/L	0 mg/L	0.000 mg/L	0.027 mg/L	0.001 mg/l	0.003 mg/L	0.00 mg/L	0 mg/L
L1X2	6.3	0 mg/L	ND	0.000 mg/L	0 mg/L	0.020 mg/L	0.213 mg/L	0.002 mg/L	0.002 mg/L	0.00 mg/L	0 mg/L
L2X3	5.6	0 mg/L	ND	0 mg/L	0 mg/L	0 mg/L	0.072 mg/L	0.003 mg/L	0.003 mg/L	0 mg/L	0 mg/L
L2X4	6.8	0 mg/L	ND	0 mg/L	0 mg/L	0.001 mg/L	0.040 mg/L	0.004 mg/L	0.002 mg/L	0 mg/L	0 mg/L

Sample	*pH*	*As 108.60*	*Be 313.10*	*Co 220.01*	*Cr 209.716*	*Cu 327.39*	*Fe 238.20*	*Mn 257.61*	*Ni 231.604*	*Pb 220.35*	*Zn 206.200*
L2Y1	–	192.5 ppm	ND	2.605 ppm	12. 34 ppm	9.785 ppm	13600 ppm	29.71 ppm	8.022 ppm	4.202 ppm	165.4 ppm
L1Y2	–	144.0 ppm	ND	2.691 ppm	4.435 ppm	7.078 ppm	8190 ppm	74.19 ppm	6.158 ppm	50.74 ppm	1272 ppm

Sample	*Fe 238.204*	*Nb 309.41*	*Sn 189.92*	*TA 226.230*
L1X1	99.5 wt per cent	0.11 wt per cent	0.30 wt per cent	0 wt per cent
L2X2	99 wt per cent	0.09 wt per cent	0.5 wt per cent	0 wt per cent

Discussion

The results of the physicochemical analysis in the above table show the concentration of the heavy metals in the ponds around the mining sites. Apart from Fe, the concentrations of the other metals investigated were negligible and sometimes not available. If the values obtained in the water samples are compared with the ore, it clearly shows that the washing process does not remove those heavy metals, an indication of complex occurrence.

Impact of Mining and Processing/Beneficiation on the Environment

Based on the above, it can be asserted that the mining of columbite in the areas investigated has minimal impact on the surrounding surface water. However, the impacts mining and processing/beneficiation pose on environment are considerable, one of such visible effects is the disturbance of land and water regime due to waste disposal. In the case of Jos, lands are not reclaimed after mining; this has deprived the inhabitants of their farmland and exposed them to danger of floods, drowning and erosion. Some of the effects of mining listed by Strahler and Strahler (1992) studied widely are observed on the Plateau. Problems that result from beneficiation are generation of solid waste, liquid waste and tailings. Records show that about 90.08 per cent of accidents are associated with mine ponds (Plate 9.10) are fatal (Adiuku-Brown, 1999).

Use of Improper Equipments

The equipments in use are outdated compared with what miners use globally, they have got no access to modern equipment to carry out mining activities, processing and beneficiation of columbite and other minerals. This leads to waste of time, excessive use of energy and poses negative effects on the quality and the yield.

Lack of Skilled Manpower

Based on the interaction with the miners and processors, it is obvious that the miners have no formal education while the processors are SSCE/GCE holders (Secondary school leavers) and so they all lack skilled manpower as well as modern equipment to carry out mining, processing and beneficiation of columbite and other minerals.

Low Wages

Miners in Nigeria and Republic of Congo earn $281 and $200 per month respectively, while their counterparts in South Africa, U.S.A Australia and Japan earn $1,385, $1,508 $1,239.31, and $1,260 respectively (www.macon.com/southafricaminers), (www.favorimineral.com) and (www.miningreview.com).

Lack of Formal Market

A centralizedl market in Africa for columbite and others that will buy at competitive prices as compared to what the vendors practice in Nigeria will go a long way to motivate miners in Nigeria and some other developing countries.

Conclusion

Based on the interaction with the miners and processors, it can be concluded that they all lack skilled manpower and modern equipment to carry out mining, processing and beneficiation of columbite and other minerals. Mechanized mining and training of manpower will enhance the yield of columbite in Nigeria which will lead to income generation.

Recommendations

It is recommended that artisanal miners in Nigeria be sent for 2–3 months exchange programme to nations under NAM – S&T that have advanced in mining such as South Africa. The experience that will be gained will be utilized in mining and processing at a small scale level. The miners should be assisted with modern equipment that will be used to carry out the mining, processing and beneficiation. Another important challenge that should be given attention is the issue of education, both the miners and those processors should be given formal education. I am advocating that NAM S&T should be the pushing force in achieving a centralized market for columbite and other minerals in Africa to motivate the miners.

Acknowledgement

The authors wish to acknowledge the financial support from the management of Raw Materials Research and Dev. Council, Abuja, Nigeria.

References

1. Adiuku-Brown, M.E. (1999). The dangers posed by the abandoned mine ponds and lotto mines on the Jos Plateau. *Journal of Environmental Sciences*. 3 (2), pp. 258–265.
2. Adiuku M.E. and Ogezi, A.E. (1991). Heavy Metal Pollution from Mining Practices: A case study of Zurak. *Journal of Mining and Geology* 8(1), pp.71–77.
3. Ajaegbu, H.I. (1985). Rural Planning and Development Prospects in a Tin mining environment: The Jos Plateau Region of Nigeria. Proceedings of the I.G.U Commission on Rural Development Neuchated, Switzerland.
4. Alkali, S.C. and Yusuf, S.N. (2010). Gravity Study over Jos-Bukuru Younger Granite Complex, North Central Nigeria [www.scholarsresearchlibrary.com (accessed 10 Sept. 2012)].
5. Eckert, J., Reichert, K., Schnitter, C. and Seyeda, H. (2001). Processing of columbite-tantalite ores and concentrates for niobium and niobium compounds in electronic applications. Symposium on Niobium [www.cbmm.com.(accessed 4 Sept. 2012)].
6. Hodder, B.W. (1959). Tin mining on the Jos Plateau of Nigeria. Economic Geography, pub. Clark University. 35 (2). pp. 109-122.
7. Macleod, W.N. Turner, D.C. and Wright, E.P. (1965). The Geology of the Jos Plateau. Vol. 1. General Geology, Geological Survey of Nigeria. Bull. No.32, pp. 118.
8. Mackleod, W.N., Turner, D.C. and Wright, E.P. (1971). The Geology of Jos Plateau, vol.1 General Geology, *Bull. Geol. Survey of Nigeria*, pp. 11–34.
9. Ndace, J.S. and Danladi, M.H. (2012). Impact of derived tin mining activities on Landuse/Landcover in Bukuru, Plateau state, Nigeria. *Journal of Sustainable Development*. 5(5), pp. 90–100.
10. Strahler, A.N. and Strahler, A. H. (1992). Environmental Geoscience, John Wiley and Sons, Inc. pp. 511.

Chapter 10

Mineral Beneficiation and Value Addition: Tanzania Perspective

Hamisi M. Komba and Elias M. Kayandabila

Mineral Beneficiation and Value Addition Section,
Ministry of Energy and Minerals, 754/33 Samora Avenue,
P.O. Box 2000, Dar Es Salaam, Tanzania
E-mail: hamiskomba@yahoo.co.uk and ekayandabila@yahoo.co.uk

ABSTRACT

This paper aims at addressing the efforts done by the Government of The United Republic of Tanzania in harnessing the mining sector reforms to ensure that mining create wealth thus, contributing significantly in the socio-economic development and therefore, poverty eradication. In an effort to promote investment in the mineral sector within the country, the government moved from the command economy towards private.

Like many other mineral resource rich developing countries, Tanzania went through major economic reforms commencing in the late 1980's that have had an impact to the minor reforms in the mining sector today. Trade liberalisation, as a result of use of good macro-economic policies set in place the first Mineral Policy of 1997 and enactment of the Mining Act of 1998. This process of policy formulation and mining code implementation was a crucial milestone in the country's mineral sector.

Today, more than 10 years of implementation of the 1997 mineral policy, priority is given to mineral beneficiation and value addition processes. Therefore, in order for the government to promote mineral processing activities within the country the ministry responsible for supervision of the mineral sector introduced a new section to oversee mineral beneficiation and value addition activities. At this stage, the section is solely responsible in coordinating interventions to unbolt the challenges and constrains faced in the Tanzania's mineral processing industry. Amongst, the current reviews made within the policy and code framework requires some minerals to be set aside for local processing and grants for processing, smelting and refinery licences.

Keywords: *Mineral value addition: Tanzania overview.*

Introduction

Tanzania is geographically located in the eastern side of the African continent and, is being bordered to eight countries namely; Kenya, Uganda, Rwanda, Burundi, Democratic Republic of Congo (DRC), Zambia, Mozambique and Malawi. However, the country is easily accessed by air, road and sea.

This Report has been prepared in connection with a 3 days International Workshop organised jointly by the Centre for Science and Technology of the Non-Aligned and Other Developing Countries (NAM S&T Centre) and Department of Science and Technology (DST), South Africa on Minerals Processing and Beneficiation. Therefore, authors of this report are focused to present information of mineral resources and utilisation potential in Tanzania, administration of the sector, different reforms made to attract mining investment, applicable laws and regulations, challenges; and opportunities available towards mineral processing in Tanzania.

Although minerals and their products plays an important role in Tanzania's economy, the mineral sector contributes only about 2.7 per cent to the GDP, which is rather low compared to the available mineral resources potential.

Mineral Resources Potential

Background

Mining has played a key role in the development of modern civilisation by supplying the structural materials for commerce, tools, jewellery, substances having various industrial uses and energy sources. Therefore, the strategic importance of mining has been both a source of development and a source of conquest and war. In his book Phillip Crowson said '*Mining is not only a means of obtaining essential raw materials, but it is first and foremost an economic activity and a source of wealth creation, just like agriculture or manufacturing*'.

The geology of Tanzania with all chrono-stratigraphical units is potential for hosting different types of earth resources, some of which are already being exploited by private companies and some are yet to be prospected and exploited. Most mineralisations fall into a number of geological environments that include Archean greenstone belts, Proterozoic Ubendian System, Karagwe – Ankolean System, proterozoic Usagaran.

Minerals found in Tanzania, as shown in the map below (Figure 10.1), include inter alia Gold, Nickel, Copper, Iron Ore, gemstone varieties (diamond, ruby, tanzanite, tsavorite/green garnet, green tourmaline, rhodolite, sapphires, emeralds, and chrysoprase) and Hydrocarbons. However, a number of industrial minerals such as kaolin, diatomite, gypsum, pozollana, limestone, meerschaum, bentonite, ball clay

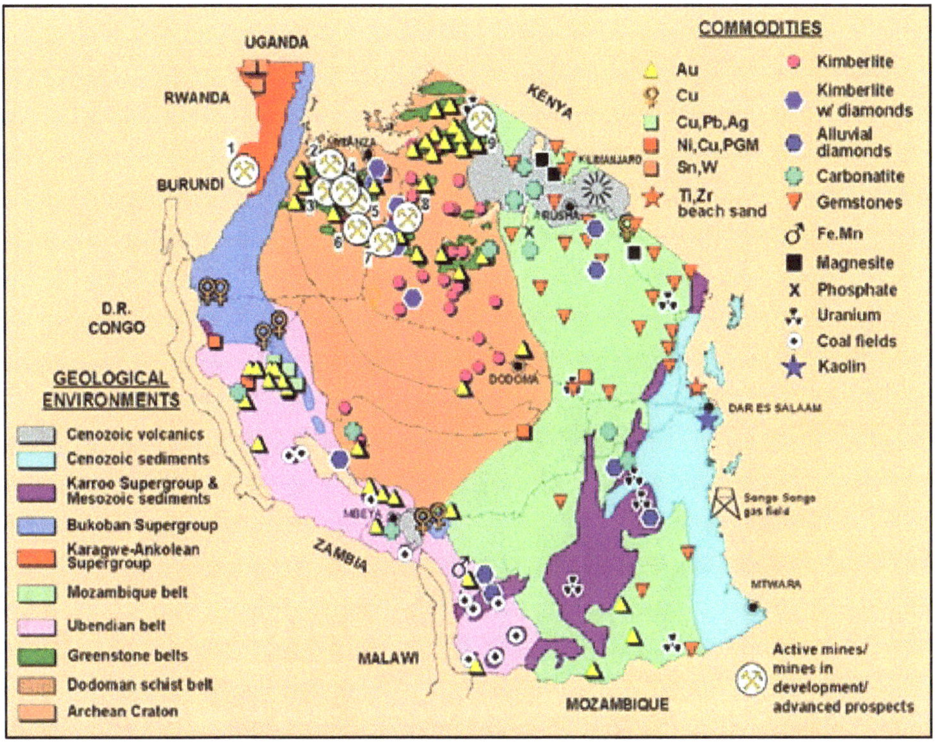

Figure 10.1: Mineral Occurrences.
(*Source*: TMAA, Annual Report 2011).

and dimension and artstones (granites, marble, anyolite) occurring in different rock formations.

Furthermore, there are six large scale gold mines (Figure 10.2) two medium scale mines (one for diamond and one for tanzanite) and a number of small scale mines mainly for gold, diamonds and coloured gemstones. Also, exploration of other minerals including uranium, coal, iron, copper and nickel is ongoing at different stages.

Mineral Reserves

Appendix 1 shows some mineral reserves for some selected minerals and projects in Tanzania. For example gold (proven, probable, indicated and measured) gold is estimated to be 31.67 million ounces, gemstones (tanzanite and diamonds) indicated and measured reserve is 87.6 million carats, nickel (inferred reserve) 1.9 billion pounds, cobalt (inferred reserve) 232 million pounds, uranium (indicated and inferred) 46.14 million pounds, coal 165.5 (proven reserve) million tons and iron ore proven reserve of 45 million tons.

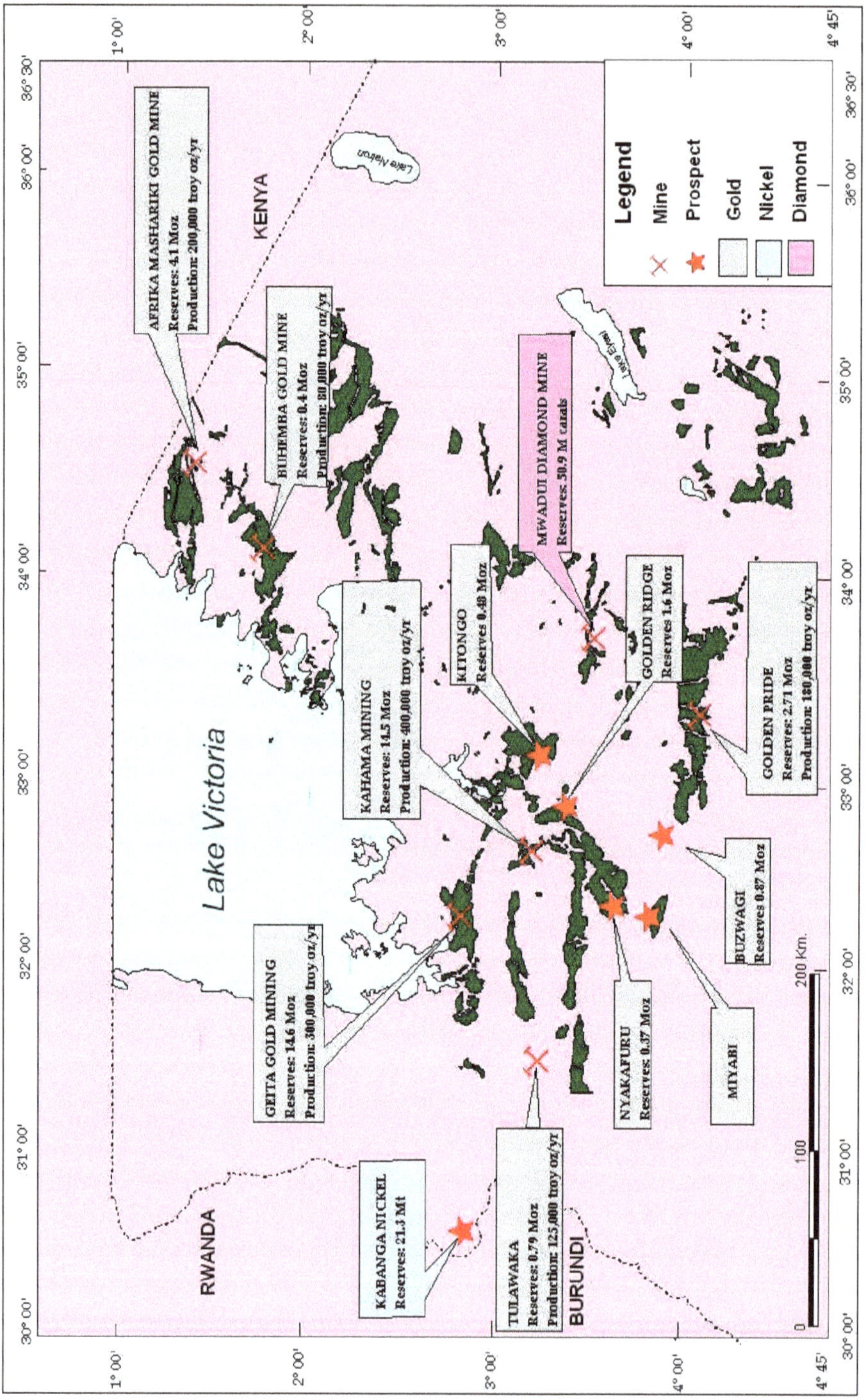

Figure 10.2: Large Scale Mines.
(*Source*: TMAA, Annual Report 2011).

Mineral Production

Appendix 2 shows minerals production based on the medium and small scale mining from the year 2005 to 2010 amounting to 22,835.72 kgs (gold), 342,366.36 Carats (diamond), 28,856,177.45 grams (tanzanite) and other industrial minerals.

Administration of the Sector

Organisation Structure

The Ministry of Energy and Minerals is mandated to facilitate development of energy and mineral resources. The mineral sector is guided by the Mineral Policy, 2009 which aims at increasing the mineral sector's contribution to the GDP and alleviate poverty by integrating the mining industry with the rest of the economy. The ministry administers the sector through The Mining Act, 2010 and Explosives Act, 1963. The Ministry is reviewing Explosives Act, 1963 and formulation of a new Mineral Beneficiation/Value Addition Act.

The minerals Division is led by the commissioner for Minerals assisted by six Assistant Commissioners for small scale mining development; mines inspection; explosives management; mineral economics and trading; mineral beneficiation and value addition; and licensing and mineral rights management. The Commissioner is also assisted by eight assistant Commissioners who are based in the eight zonal offices as it was found necessary to bring extension services close to the people. Again for the purpose of bringing services related to the development of the mining sector in the country close to the public, Resident Mines Offices are established at areas where there is a lot of mining operations and, are usually located at the district's headquarter.

Establishment of Mineral Beneficiation and Value Addition Section

Mineral beneficiation and value addition activities that include gemstone cutting, polishing and jewellery as well as metal smelting and refinery enhance the quality and value of minerals. Mineral beneficiation and value addition improves local skills in the mining industry; creates employment; increases income and contributes to the national economy. These benefits can be realized if and only if mineral beneficiation and value addition is undertaken within the country.

To be able to carry out mineral beneficiation and value addition activities, it is important to have the necessary technical skills, proper processing technology for specific ores, access to capital and adequate infrastructure and facilities. In this regard the Mineral Policy of Tanzania of 2009 encourages the Government to promote investments in mineral beneficiation and value addition such as lapidary, stone carving and jeweler making; smelting and refining industries.

In order to promote mineral processing the Ministry of Energy and Minerals established a new section to oversee mineral processing within the country. This section coordinates monitoring and regulation of mineral beneficiation activities that include mineral processing, smelting and refinery. It also facilitates investments and promotion of mineral beneficiation and value addition such as lapidary, stone carving and jewellery making, smelting and refining industries. The Section is also responsible for processing license applications for processing, smelting and refining of minerals.

Different Reforms Made to Attract Mining Investment

Evolution of International Mineral Investment Law and Policy

The laws and policies applied to international mining investments in developing countries can be separated into three eras namely (i) pre-1960 (colonial era) period (ii) 1960 – 1985 (post-colonial era) period and, (iii) the post-1985 years (modern era). Each of these eras reflected a unique view guided by an accepted economic theory and development paradigm.

The in-depth explanation of the three different periods is not the purpose of this paper and, therefore, it requires another platform. However, this paper aims to highlight the major events of the third era in relation to Tanzania as follows:

- During post-1985 years (modern era) the most dramatic change is reflected in the effort to privatize mining investments. Thus, mineral rich developing countries established an enabling environment for mining by reforming their mining legislation, capacity building of government supervisory institutions for the mining sector, increasing the availability and reliability of geological information, creating competitive fiscal regimes and promoting investment opportunities.

The Genesis of Economic Reforms and Investment Promotion in Minerals Sector

Major economic reforms had a direct impact on the minor economic reforms that occurred in the Tanzanian mineral sectors, adoption of the use of good macroeconomic policy that led the government from command economy to private.

Apart from lack of skilled labour and experienced persons to manage the mines, the states owned companies that were given the mandate to operate the mines failed because of the difficulties in accessing foreign money to obtain and maintain the mining equipments. Moreover, all major sectors of production were inactive and hence forced the state to revise her economic policies.

Besides, for the state to create the investors' confidence, an investment Act of 1997 introduced. Among other things, the act promoted private investment and guaranteed the investors against nationalization. In addition, the investment act of 1997 assured fiscal stability to the investors. Knowing the nature of the extractive industry, the government saw the need to introduce a separate fiscal regime and taxation package within which the investors might recover their respective costs for investment.

In the 1990s Tanzania, a mineral rich country started to deregulate mining codes and policies, further more they promoted investment policies. However, the trend in the use of good macroeconomic policies were emphasised where by the state attracted the private ownership of investments, this was the crucial step for the government to convince private investments into mining projects.

The new mineral taxation was set in place in 1997 under the Financial Laws (Miscellaneous Amendments) Act of 1997, only royalty is being charged under the mining act. Table 10.1 Show fiscal regimes and taxation system and the percentage.

Table 10.1: Regimes and Taxation System and the Percentage

Corporate tax	30 per cent
Royalty (based on net value)	3 per cent for all minerals, 5 per cent for diamonds only
Tax holiday	none
Capital allowance	100 per cent
Withholding Tax on technical services	5 per cent (local) and 15 per cent (Foreigners)
Withholding Tax on dividends	10 per cent
Withholding Tax on interest	10 per cent
No capital gain tax	

Under the current Mining Act, 2010;

Royalty (based on gross value) Uranium 5 per cent, Gemstone and Diamond 5 per cent, Metallic Minerals (copper, gold, silver and platinum group minerals) 4 per cent and Other Minerals (building materials, salt, all minerals within industrial minerals group) 3 per cent

Among other things the government of Tanzania introduced some investment promotion under Tanzania Investment Centre (TIC), including the mineral sector. Therefore, in the 1990s the government came up with an Investment Act which then was amended in 1997 which guarantees private investors on the fiscal regime stability (Investment Act 1997, Section 19), freely transferability of the profits and dividends (Investment Act 1997, Section 21) and most important was to return investor's confidence against nationalisation or expropriation (Investment Act 1997, Section 22). However, all investors with an incentive certificate issued by Tanzania Investment Act of 1997 are eligible to deduct costs incurred with respect to mining activities as soon as revenues are realised (operation costs as defined by the Act).

Thus, it is from this background that the Government of the URT opted for the structural adjustment programme (SAP) in the late 1980's as engineered by development partners including the International Financial Organisations. Amongst, the mining industry economic reforms were the result of the major reforms that had started in 80's. Following the considerable reforms the key milestone was the implementation of the first Mineral Policy, 1997 and a competitively Mining Act, 1998. The formulated mineral policy adopted the use of macroeconomic policy where by mining investments were left to private sector mainly, the Foreign Direct Investors. Consequently, mining sector expanded rapidly and today Tanzania is amongst the four top African gold producers (Figure 10.3).

Mineral Policy and Mining Code in the Context of Mineral Beneficiation and Value Addition

After 10 years of implementation of the first and newly formulated mineral policy of 1997, the government recognised the need for reviewing the policy. This was due to

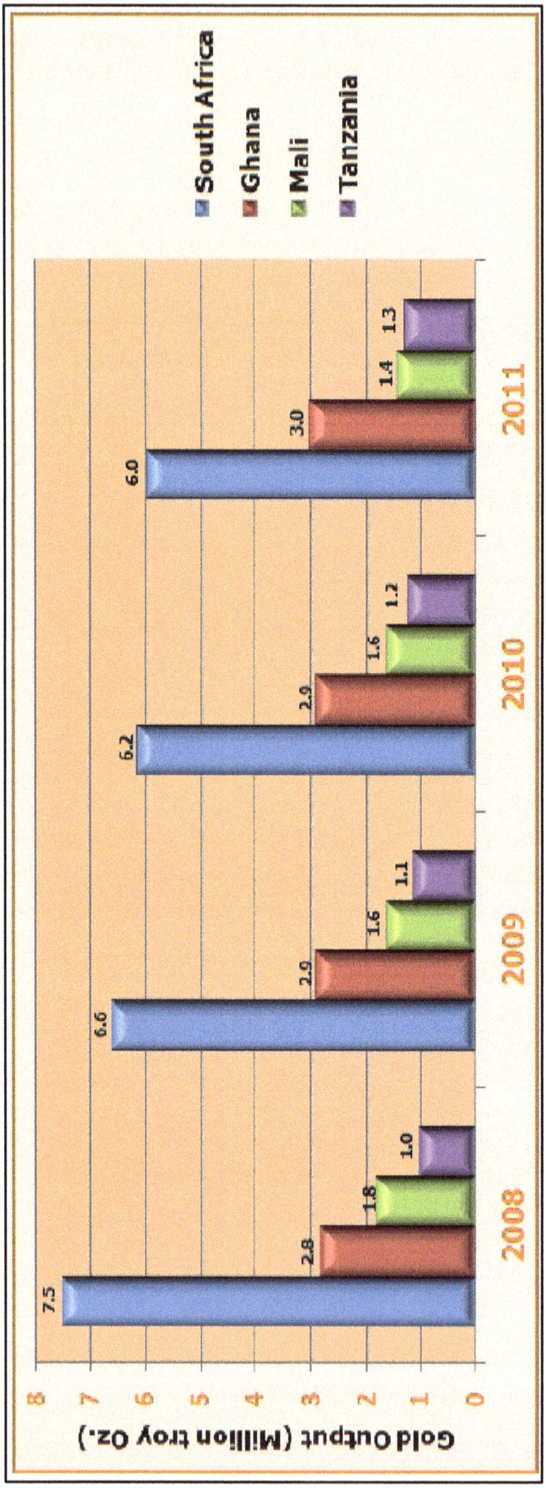

Figure 10.3: Africa's Top Four Gold Producing Countries (2008–2011).
(*Source*: TMAA, Annual Report 2011).

the fact that the government liaised with the voice of the local people against mining companies who were seen to operate in an enclave and, no economic spin-off effects observed. During this period a number of committees were formed to advice on how best the mineral resources can be managed. Different committees formed led by the following people namely Dr. Kipokola, Honourable Masha, Dr. Bukuku, Rtd. Jorge Bomani and Rtd. Gen. Mboma. Therefore, the findings from their reports resulted in putting in a place the current Mineral Policy of 2009 and, thereafter the Mining Act of 2010.

Mineral Policy, 2009

The main features of the mineral policy 2009 are as follows:

Vision

"Effective mineral sector, contributing significantly to the acceleration of socio- economic development through sustainable development and utilisation of mineral resources in Tanzania by 2025."

Mission

"To set policies, strategies and laws; regulate mineral exploration, production, trading, value addition and mineral actors for sustainable development of mineral resources; and integrate with other sectors of the economy."

Among others, the government's mineral policy objective aim at facilitating, supporting and promoting increased participation of Tanzanians in gemstone mining, value addition activities within the country to increase income and employment opportunities.

In this regard, it is worthwhile to define the mineral beneficiation and value addition activities. Different jurisdictions have got different meaning about beneficiation and value addition activities. Practically the words Beneficiation and Value Addition are used interchangeably to refer to.

In view of the above the Tanzania Mineral Policy of 2009 says that mineral beneficiation and value addition activities will include gemstone cutting, polishing and jewellery as well as metal smelting and refinery. Moreover, this policy requires the government to promote investment in the fabrication, manufacturing sectors, lapidary, stone carving and jewellery making and to work in partnership with the private sector, regional and international organisations to strategically invest in smelting and refining industries.

The Mining Act, 2010

The mining code of Tanzania provides for applications and grants for processing, smelting and refinery licences regardless whether the applicant has or has not the mineral right. However, the code requires concession owners to set aside some minerals for the local processing, smelting or refining. The amount of minerals to set aside will be determined by the Minister after a consultation with concession owners and the Mining Advisory Board (MAB).

The mineral beneficiation and value addition licensing procedure is centralised, all mineral processing application are channelled to the Commissioner for Minerals at the ministry's headquarters. At this juncture, the Commissioner if satisfied with the content of application will only issue a processing license, whilst the application for smelting or refining license will be forwarded to the Minister for issuing.

In the effort to encourage value addition activities within the country the government had set different royalty rates for rough and cut gemstones. When exporting rough gemstones the licensee will have to pay 5 per cent of the gross value of the minerals meanwhile, the cut gemstones is 1 per cent of the gross value.

Challenges and Constraints for Investment and Development of the Mineral Processing Industries

Mineral Processing for Value Addition

Most of minerals, except building materials, produced in the country are exported without being value added owing inadequate capacity to do so. There are deliberate efforts to promote carrying out of value addition of the minerals before being exported in order to maximize benefits accruing from development of mineral resources. However, industrial minerals produced are used directly in industries to produce various products which include cement.

Insufficient Power Supply

Tanzania's total installed power generating capacity is around 1438MW, of which is hydropower contributes 561MW, thermal 877MW and wind 0MW. The country is actually experiencing an energy shortage and there is a lack of electric power in areas with mineral resources which is not encouraging mining and processing investments. The volume of mineral beneficiation and value addition activities require adequate and sustainable power supply.

Availability of Raw Materials for Local Beneficiation

The current set up in the mining industry gears toward export orientation of raw material, with large scale mining companies being tied into long term contractual obligations with their overseas clients, refiners and smelters.

Inadequate Knowledge for Developing Local Beneficiation and Value Addition Required

Tanzania seeks for enough specific attention to scientists to unlock the problems facing in terms of research and development to sufficiently service the local beneficiation

Poor Infrastructure

Mining operations are conducted in remote areas with very poor road networks to established hubs in relations to connecting the two.

Gemstone Cutting and Polishing, Rock and Mineral Carvings and Jewellery Fabrication

The gemstones value addition requires the establishment of integrated Jewellery Hubs throughout the country.

Solution

To unlock the above illustrated challenges and constraints for investment and development of the mineral beneficiation and value addition programmes the government of Tanzania requires coordinating interventions from different key stakeholders to either remove or significantly reduce the said stumbling blocks. In the due course, it is therefore paramount for the government to prepare a road map and engage all mineral beneficiation stakeholders to respond to each of the above constraints and proposes the intervention.

It is vital for the government to make sure necessary infrastructures to support beneficiation is properly developed; invest in the human capital development; and support and develop technologies by research and development.

Conclusion

The United Republic of Tanzania offers the following competitive advantages:

- ☆ Considerable unexploited mineral potentials.
- ☆ Accessible geological data and information.
- ☆ An attractive investment climate for both local and foreign investors (Policy, Legal and Regulatory Framework conducive to private sector participation *e.g.*, The National PPP Policy, 2009 and PPP Act, 2010 are in place).
- ☆ The Government is committed to attracting private investment in the mineral sectors, mineral beneficiation and value addition in particular.
- ☆ Transparent process of granting mineral rights; and
- ☆ Above all, she has Peace and political stability.

Acknowledgements

On behalf of the Ministry of Energy and Minerals, the authors would like to express their sincere appreciation to the Centre for Science and Technology of the Non-Aligned and Other Developing Countries (NAM S&T Centre) and Department of Science and Technology (DST), South Africa for the two-fold; First is for organizing such an important workshop to impart knowledge and exchange experience; and Second, is for the financial support.

List of Abbreviations

DRC: Democratic Republic of Congo
URT: United Republic of Tanzania
SAP: Structural Adjustment Program
WB: World Bank

IMF: International Monetary Fund
TIC: Tanzania Investment Centre
GDP: Gross Domestic Product
RTD: Retired
LBMA: London Bullion Market Association

References

1. Ministry of Energy and Minerals, the Mineral Policy of Tanzania (Dar es Salaam, Tanzania: Tanzania Printers, 1997).
2. Ministry of Energy and Minerals, The Mineral Policy of Tanzania (Dar es Salaam, Tanzania: Tanzania Printers, 2009).
3. Ministry of Energy and Minerals, Opportunities for mineral resource development (Fourth Edition) (Dar es Salaam, Tanzania: Tanzania Printers, 2005).
4. Otto James and Cordes John, The Regulation of Mineral Enterprises: A Global Perspective on Economic, Law and Policy (Colorado, United States of America: Rocky Mountain Mineral Law Foundation, 2002).

Internet

5. MINISTRY OF ENERGY AND MINERALS http://www.mem.go.tz/(last visited 03 September, 2012)
6. Citigroup Global Markets Inc. Tanzania: A decade of solid growth, but further reforms and challenges https://fidirect.citigroup.com/(last visited 30 August, 2012)
7. Financial Laws (miscellaneous amendment) Act, 1997 http://www.iornet.com/newiornet/cp/tanzania23.htm last visited 31 August, 2012)
8. Godwin, M. H. N., The Impact of Changes in Government Policies on the Development of Mineral Resources in Tanzania (2000) http://www.dundee.ac.uk/cepmlp/car/html/car4_art3.htm (last visited on 31 August, 2012)
9. Tanzania investment Act 1997 http://www.iornet.com/newiornet/cp/tanzania23.htm (last visited 31 August, 2012)
10. Tanzania Investment Centre: Investment Incentives http://www.tic.co.tz/(last visited 30 August, 2012)
11. Roger Baxter, Mineral Beneficiation: The Mining Industry Viewpoint (2005) http://www.lbma.org.uk/assets/2c_baxter_lbma2005.pdf (last visited 3 September, 2012)
12. Phillip Crowson, Old wine in new bottles: Policy Issues for the Mining Industry http://books.google.co.tz/books?id=CbBNvKVNrjgC&pg=PA611&lpg=PA611&dq=%E2%80%98Mining+is+not+only+a+means+of+ obtaining+essential+raw+materials,+but+it+is+first+and+foremost+an+ econom ic+activity+and+a+source +of+wealth+creation,+just+lik e+agriculture+and+manufac

turing%E2%80%99&source= bl&ots=TMWuue3EOw&sig=bJ R7hzOidwr4ifs5I SvMmsbOfY&hl=sw#v=onepa ge&q=%E2%80%98Mining%20is%20not%20only%20a%20means%20of%20obtaining%20essential%20raw%20materials%2C%20but%20it%20is%20first%20 and%20foremost%20an%20economic%20acti vity%20and%20a%2 0source%20of%20wealth%20 creation%2C%20just%20like% 20agriculture% 20and%20manufacturing%E2% 80%99&f=false (last visited 5 September, 2012).

13. Tanzania Minerals Auditing Agency, TMAA Annual Report 2011: http://www.tmaa.go.tz/(last visited 5 September, 2012.

Appendix 1

Sl.No.	Mine/Project Resources/Ore Reserves	Owner	Location	Mineral	Mineral	Remarks
1.	Bulyanhulu Gold Mine	African Barrick Gold (100%)	Kahama	Gold	11 million ounces	Proven and probable
2.	Buzwagi Gold Mine	African Barrick Gold (100%)	Kahama	Gold	2.9 million ounces	Proven and probable
3.	Geita Gold Mine Ashanti Limited	Anglo Gold (100%)	Geita	Gold	8.73 million ounces	Indicated
4.	Golden Pride Mine Limited	Resolute Mining (100%)	Nzega	Gold	0.16 million ounces	Proven and probable
5.	North Mara Gold Mine	African Barrick Gold (100%)	Tarime	Gold	2.8 million ounces	Proven and probable
6.	Tulawaka Gold Mine and Northern Mining Exploration Ltd (30%)	African Barrick Gold (70%)	Biharamulo	Gold	67,000 troy ounces	Proven and probable
7.	Golden Ridge Gold Project	African Barrick Gold (100%)	Shinyanga	Gold	0.53 million ounces	Indicated
8.	Nyanzaga Gold Project	African Barrick Gold (100%)	Sengerema	Gold	0.3 million ounces	Indicated
9.	Handeni Gold Canaco Project	Handeni Resources Inc. (100%)		Gold	3.94 million ounces	Measured and Indicated
10.	Chunya Gold Project (New Luika Gold Mine)	Shanta Mining Limited (100%)	Chunya	Gold	0.36 million ounces	Indicated
11.	Buckreef Gold Project	Tanzania Royalty (55%) and STAMICO (45%)	Geita	Gold	0.95 million ounces	Indicated
12.	TanzaniteOne Mine	TanzaniteOne Mining Limited (100%)	Simanjiro	Tanzanite	63 - 83 million carats	Indicated
13.	Williamson Diamonds Mine	Petra Diamonds Ltd. (75%) and URT (25%)	Kishapu	Diamonds	4.6 million carats	Measured and Indicated
14.	Kabanga Nickel Project	Barrick Gold Corp. (50%) and Xstrata Plc (50%)	Ngara	Nickel	1.16 billion pounds	Inferred
15.	Dutwa Nickel Project	African Eagle Resources Plc. (90%)	Mwanza	Nickel Cobalt	751.8 million pounds 23.2 million pounds	Inferred Inferred

Contd...

Appendix 1–Contd...

Sl.No.	Mine/Project Resources/Ore Reserves	Owner	Location	Mineral	Mineral	Remarks
16.	Mkuju River Project	Uranium One Inc. (100%)	Namtumbo	Uranium	28.5 million pounds	Indicated
17.	Manyoni Uranium Project	Uranex (100%)	Bahi/ Manyoni	Uranium	17.64 million pounds	Inferred
18.	Mchuchuma Coal Mine Corporation	National Development	Ludewa	Coal	125.3 million tons	Proven
19.	Mbalawala Coal Project/ Ngaka Basin	Intra Energy Corporation (70%) and NDC (30%)	Ruvuma	Coal	40.2 million tons	Proven
20.	Liganga Iron Ore Project	National Development Corporation, 2011	Ludewa	Iron Ore	45 million tons	Proven

Source: TMAA, Annual Report, 2011.

Appendix 2: Minerals Produced by Medium and Small Scale Miners (2005–2010)

Mineral	Unit	2005	2006	2007	2008	2009	2010	TOTAL
Gold	Weight (Kg)	1,019.55	5,163.99	4,754.86	4,467.74	5,228.11	2,201.47	22,835.72
	Value (USD)	13,802,570.80	95,136,100.99	113,966,372.25	104,136,437.23	123,703,657.67	71,345,177.92	522,090,316.86
Diamond	Weight (Carat)	24,498.05	84,930.86	70,278.97	87,321.44	46,435.53	28,901.51	342,366.36
	Value (USD)	3,100,704.69	5,396,273.85	6,814,623.09	6,309,293.95	5,759,160.96	4,755,561.52	32,135,618.06
Tanzanite	Weight (g)	1,306,655.22	5,504,055.46	6,299,513.89	5,030,610.74	5,012,484.39	5,702,857.75	28,856,177.45
	Value (USD)	34,016,728.35	20,757,607.91	29,200,927.31	22,103,371.28	9,091,121.20	17,777,725.67	132,947,481.72
Limestone	Weight (g)	102,100.00	109,086.40	164,543.90	207,922.45	106,185.00	120,000.00	809,837.75
	Value (TZS)	6,977,547.15	7,455,000.00	10,908,400.00	15,941,900.00	19,363,500.00	21,882,751.80	82,529,098.95
Galena	Weight (Kg)	–	–	83.40	248.50	20.50	109.00	461.40
	Value (TZS)	–	–	37,503,488.00	133,000,000.00	1,650,000.00	12,450,500.00	184,603,988.00
Coal	Weight (Tonne)	–	–	2,376.30	1,316.70	–	–	3,692.30
	Value (TZS)	–	–	68,912,555.00	38,756,612.00	–	–	107,669,167.00
	Weight (Tonne)	7,401.86	5,500.00	–	2,120.00	1,820.00	–	16,841.86
Gypsum	Value (TZS)	218,361,396.00	162,254,840.00	–	62,541,865.60	64,676,000.00	–	507,834,101.58
Zircon	Weight (Tonne)	6,732.32	7,847.43	–	–	–	–	14,579.75
	Value (TZS)	1,307,875.00	1,955,000.00	–	–	–	–	3,262,875.00
Tin	Weight (Tonne)	2,035.00	2,060.00	1,220.50	–	1,000.00	–	6,315.50
	Value (TZS)	–	–	9,463,000.00	–	5,853,540.00	–	15,316,540.00
Kaolin	Weight (Tonne)	–	–	1,015.00	28,896.00	18,624.00	48,535.00	
	Value (TZS)	–	–	2,500,000.00	28,896,000.00	18,624,000.00	–	50,020,000.00
Copper	Weight (Tonne)	–	–	–	750,000.00	3,000,000.00	3,750,000.00	
	Value (USD)	–	–	–	22,500.00	–	150,000.00	172,500.00

Chapter 11

Processing Uganda's Mineral Resources 2012

William Balu-Tabaaro

Mineral Dressing Engineer/Director,
Geo-engineering Technologies Ltd. P.O. Box 654 Entebbe, Uganda
E-mail: eng_balutabaro@yahoo.com

ABSTRACT

Many countries are endowed with mineral resources but due to lack of expert knowledge in mineral extraction technologies the minerals are only processed without adding much value. As a result, the mineral concentrates are exported at low cost and these countries end up importing products made from these minerals at high cost. Lack of expertise is a result of former colonial policies that hindered growth of local skills as all the resources were shipped abroad to satisfy the colonial empires. At this critical stage of development the Non-Allied Movement (NAM) countries need to invest in training of scientists and technologists in the mineral processing of their mineral resources. They also need to train in skills for upgrading and adding value to these resources so that products can be made locally and jobs can be created. There is further need for greater technical cooperation among the Nam countries in transfer of technology and investment since some Nam countries have acquired skills and technologies in this field

Keywords: *Mineral potential, Exploitation, Skills and expertise, Value-addition, Conducive policies, Challenges and opportunities.*

Introduction

Uganda is located in the eastern part of Africa between latitudes 110000S to 440000N and longitudes 170000E to 610000E covering 236,040 sq. km, 199,710 and 36,330 sq. km land and water respectively, with a population of 31,367,972(July

2008(UN Est). The country is endowed with a wide variety of both metallic and industrial minerals, exploitation of which contributes to less than 0.7 per cent of the of the country's gross domestic product (GDP). These minerals include: beryl, bismuth, cobalt, columbite tantalite, coltan, copper, chromite, gold, iron ore, tin (cassiterite), wolfram, asbestos, clay, diatomite, feldspar, granite gnesiss, graphite, gypsum, kaolin, kyanite, limestone, marble, mica, phosphates, rock salt, silica sand, talc, vermiculite, etc. The country also has potential occurrences of uranium, diamonds and other gemstones.

Exploitation of these mineral resources range from artisanal, small-scale to large scale. Artisanal operations are those that apply rudimentary methods and are usually not very legal. Meanwhile small-scale operations are those which initially invest less than about Ush. 10 million on an operation and do not employ "specialized technology" (1 US$ = Ush. 2470(2012), and large scale operations are those with more capital investment and highly mechanized, all being legalized operations with location license and mining leases respectively.

Artisanal and small-scale mining (ASM) in Uganda employs about 130,000 men and women miners in gold, cassiterite, limestone, salt, marble, gypsum, clay works, sand and quarrying etc. The sub-sector is estimated to produce more than 90 per cent of most minerals, the biggest workforce of which being in Industrial and Building Minerals sub-sector *(Jennifer Hinton, 2006).* The ASM sector constitutes of miners; small and formal companies with some degree of mechanization, permanent miners with low-tech methods and somewhat organised, seasonal miners who often migrate from site-to-site and apply low-tech methods, rush miners, and shock-push miners.

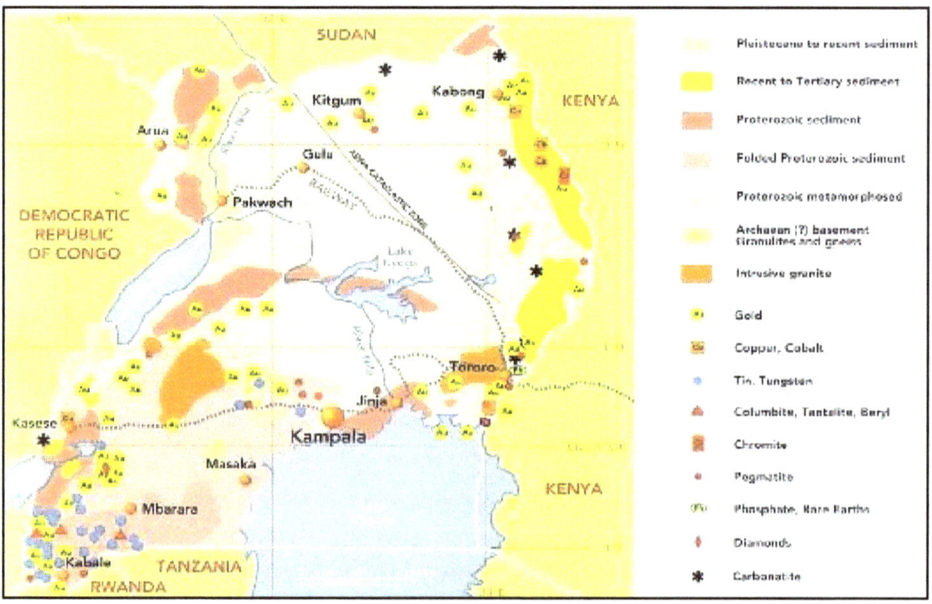

Figure 11.1: Mineral Occurrences of Uganda.

Table 11.1: Uganda's Mineral Production Returns over the Past Year (July 2006 to June 2007)

Mineral	*Returns in Tons*			
	Jul–Sept	*Oct–Dec*	*Jan–Mar*	*Apr–Jun*
Limestone	40,322.60	51,571.00	96,973.00	117,487.00
Pozollana	41,560.00	29,844.00	52,357.30	57,718.90
Gold	0.00	0.01	0.01	0.01
Vermiculite	546.00	538.00	2,609.00	0.00
Cobalt	173.71	174.81	58.71	109.60
Wolfram	33.40	4.28	19.91	7,650.00
Syenitic aggregate	2,772.00	702.00	473.00	1,958.00
Kaolin	0.00	0.00	1,596.48	1,783.49
Iron ore	0.00	0.00	194.76	179.79
Gypsum	0.00	0.00	39.48	8.00
Lead	0.00	3.00	0.00	38.00
Coltan	0.00	0.01	0.00	0.00

Note: Discrepancies in returns for some of the commodities are due to failure of production companies to submit returns or production stoppages.

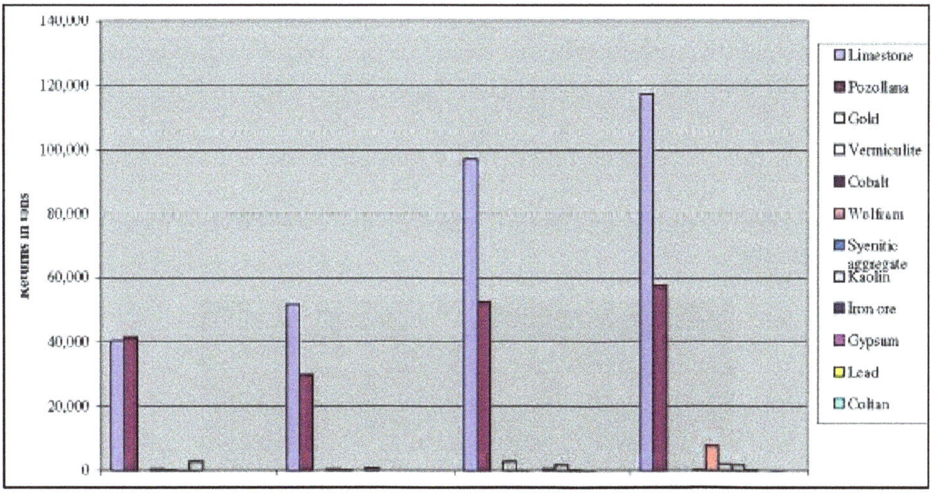

Figure 11.2: Mineral Production Returns from July 2006 to June 2007.

Processing Methods in Application

Processing methods vary from one mineral commodity to another. However variations also exist within mineral commodities depending on occurrences and scale of operation and location.

Gold

Gold processing techniques in Uganda are dependent on whether alluvial or hardrock ore is being processed and levels of capital investment, that is, artisanal to large scale variations.

Artisanal Gold Processing Technologies used by artisanal gold miners range from concentration of gold ore using pans along water streams to use of sluices. Most miners simply dig weathered ore and haul it to a stream or an improvised pond to be panned or simply scoop stream sediments and pan. Here no further size reduction is done.

Where sluices are in use the ore is feed to the sluice using shovels and pans, and water delivered either using a pump or basins from a nearby stream or pond.

However where hard rock is mined, ore is dug using pick axes and shovels, then crushed using hammers and mortars and pestles. The crushed ore is then concentrated using sluices and the concentrate amalgamated to extract gold contained in it. Most of these artisanal operations are yet to adopt good mercury use practices.

Small scale Gold Processing Small scale gold mining operations are mechanized to a great extent and continuously improve on their recovery techniques with technical support from the Government Department mandated to sustainable manage the utilization of mineral resources. One such operation, Kisita Gold Mine has demonstrated abilities of becoming a recognizable gold mining force in the country. The mine has a process plant constituting of 10 ton per hour primary and secondary crushing units which crush to minus 5″ and 1″ respectively. Crushed ore is stored in a bin before being conveyed to 6 tons per hour rod mill, capable of grinding to minus 5 mm.

Currently the mill is operating dry at 4 tons per hour, with milled ore at the discharge being washed into sluices where some gold nuggets are recovered before proceeding to Bowl Centrifugal Concentrators for final concentration (refer to Figure 11.4 for schematic).

Plate 11.1: Sluices at Artisanal Operations in Bushenyi District.

Plate 11.2: Sluices at Artisanal Operations in Amonikikene, Busia District.

Plate 11.3: Ball Mill that Failed to Run (Left) beside the Functioning Rod Mill (Right).

The mine is installing a new plant to have a 2 mm screen after the sluices to provide size control for the mill before the ore is concentrated using a Gemini Table followed by a Falcon Concentrator. The sluices are also to be redesigned into one compact vertical zig-zag unit with optimized parameters (inclination and flow rate) for improved recovery to replace the ones that are arranged in the present series set-up in Plate 11.4 (refer to Figure 11.3 for schematic). However, the company was also advised to install a hydro cyclone for size control of feed to the Falcon Concentrator other than a screen as it is an appropriate classification equipment for the minus 2 mm feed requirement. The hydrocyclone would also provide better flow optimization.

Plate 11.4: Water Spraying Milled Ore from the Rod Mill onto a Sluice.

Tailings Management: Tailings are currently being stockpiled with an improvised drainage for effluent to go down into a valley where it is pumped back to the plant for reuse. The mine however plans to use waste rock stockpiled to build a tailings dam nearby. The design and construction of the tailings needs to take into

Plate 11.5: Bowl Centrifugal Concentrators (Photo by C. Lubangakene).

consideration the fact that the mine intends to setup a chemical plant in the near future.

An alternative process at the mine is the pilot plant which handles overburden containing recoverable gold. The pilot plant has a capacity of 3 ton per hour but has reached the end of its life. The process involves a backhoe feeding a receiver bin from which the ore is conveyed through a tumbler at the end of which is a 5 mm screen. Here the ore is washed and the gravel (pebbles, cobbles or boulders) which are oversize, are stockpiled. The minus 5 mm ore proceeds to two sluice boxes where the first concentration is carried out. The runoff from the sluice boxes proceeds to a pulsating duplex jig where the next concentration is done. From the jig the concentrate is taken to a shaking table for final concentration. Tailings from the shaking table are being stockpiled in a tailings dam for further processing using chemical method. A Gemini table is used to further concentrate the shaking table concentrate as its operations are also tested (refer to Figure 11.6 for schematic) for the new plant.

Plate 11.6: Tailings from New Plant (Photo by I. Ntujju).

Concentrate from the two plants are then fed into an amalgamation barrel. From the amalgamation barrel the material is panned so as to separate the amalgam from the tailings. The amalgam is burned in a retort to extract gold from mercury.

Areas for Improvement

The mine is yet to build a separate amalgamation tailings containment as at present they are being discharged together with tailings from the jigging operatio,which is anenvironmental catastrophe,and even worse cynidatio of the tailings is carried out.

Plate 11.7: Shaking Tables (Gemini Table in foreground) (Photo by I. Ntujju).

Overburden jigging operation, which is an environmental catastrophe, and Tailings even worse when cyanidation of the tailings is carried out.

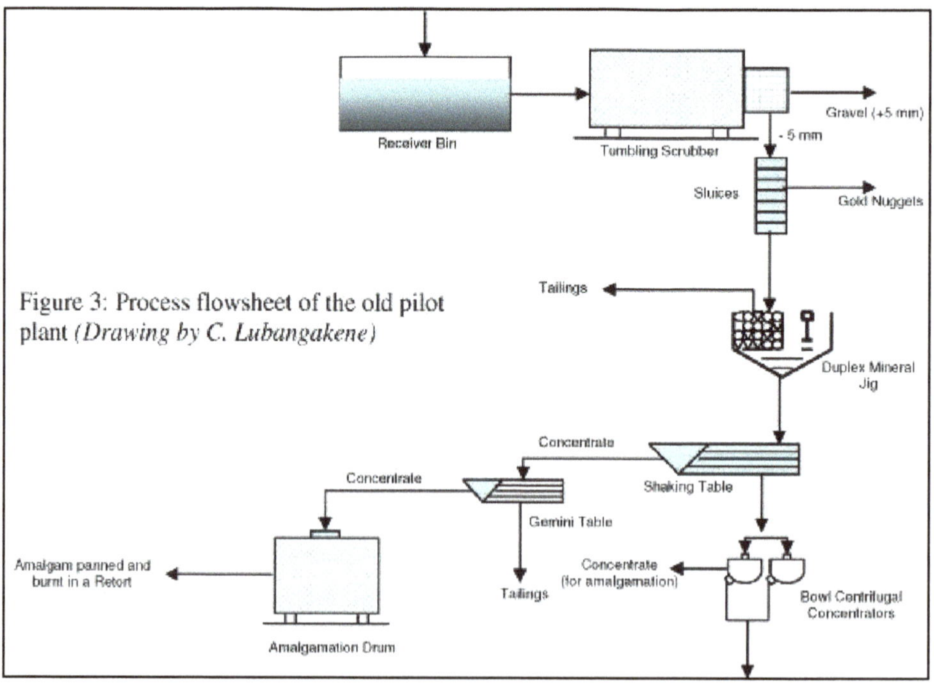

Figure 11.3: Process Flowsheet of the Old Pilot Plant (Drawing by C.Lubangakene).

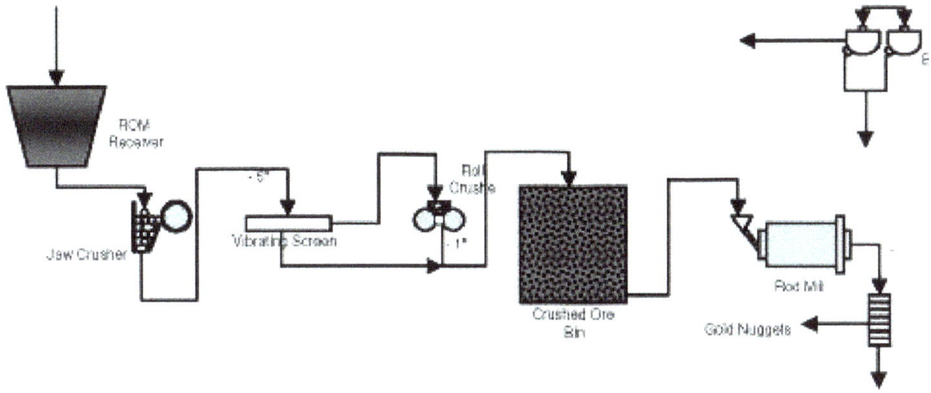

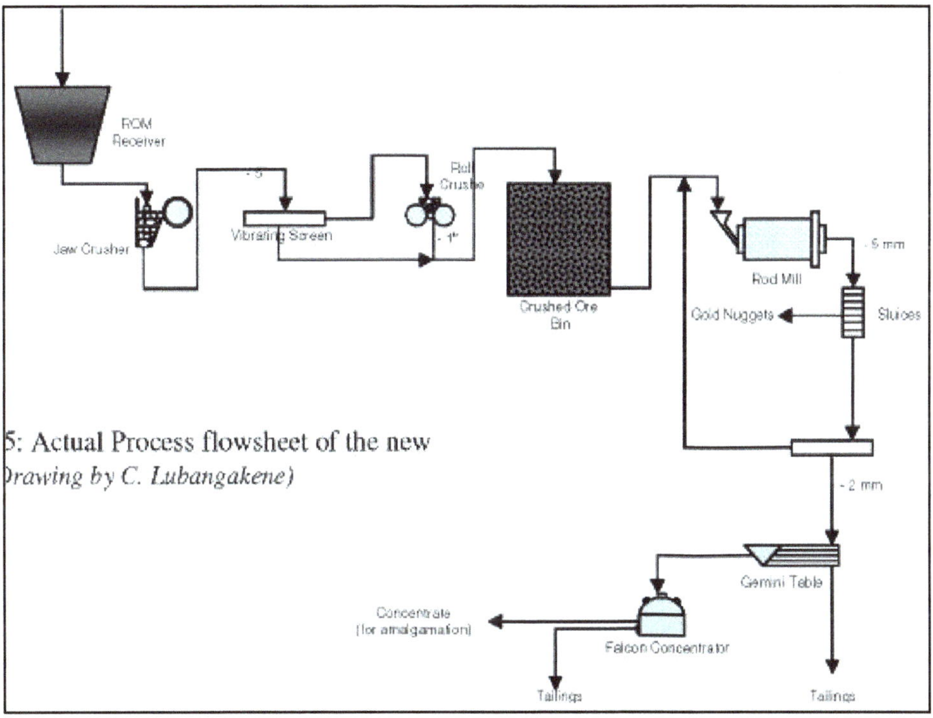

Figure 11.4: Process Flow Sheet of the New Plant being Test Run ((Drawing by C. Lubangakene).

Large Scale Gold Processing The only large scale gold mine in Uganda is the Tiira Gold Mine. The mine is both open pit and underground with an ore grade range from 10 to 20 g/t. The mine has a process plant with a recovery of 95 per cent, utilizing both gravity separation and cyanidation. Ore is crushed in three (3) stages to a product size of minus 8 mm before being milled to minus 75 microns (size control being provided by hydrocyclones) for cyanidation. Underflow of the hydrocyclones is concentrated using a Knelson concentrator and the concentrate amalgamated using

Plate 11.8: Process Plant at Tiira Gold Mine.

Plate 11.9: Tailings at Tiira Gold Mine.

a retort, refined, and then smelted. The overflow however proceeds for cyanidation, elution, electrolysis, refining and finally smelting.

Tailings Management

Tailings from the plant is pumped to the tailings dam (~ 100 m × 50 m × 6 m) and effluent collected through trenches around the dam to settlement tanks before being pumped back to the plant.

Plate 11.10: Settling Tanks.

Lead

Lead mining currently is not mechanized. The active open cast operation uses shovels, hoes, pick axes, screens and pans. However with more capital input the operation intends to open the previous underground adits and use sluice boxes, rotary pans and jigs in its processing.

Tin

Tin mining operations currently are on artisanal levels of mechanization, although with some of the operations legalized. The mining processes involve use of pick axes, shovels, jack hammers, crowbars and pans. Mined ore is hauled to nearby streams where they are panned. However, hand sorting is also carried out especially on the tailings stockpiles worked in the past.

Wolfram

Wolfram production in Uganda is currently at artisanal and slightly higher and more mechanized (especially in crushing) technological levels. At artisanal operations, picks axes, crowbars, spades, chisels, hammers, hoes and pans are used. Therefore the processing is basically using hammers to crush the ore and hauling crushed ore to streams or ponds to concentrate.

Plate 11.11: Tin Miners in Ntungamo District.

Plate 11.12: A Miner Panning Wolframite Ore in a Pond a Long a Creek at Kirwa Wolfram Mine (Photos by C. Lubangakene).

Meanwhile in the slightly more mechanized operations, processing involves primary crushing to minus 50 mm followed by secondary crushing to minus 5 mm. The size reduction operation is followed by classification using five (5) sieves into five (5) categories of material. The first category of material which is the smallest sieve size is panned in a pond or trough. After panning the concentrate is dried in an improvised roasting unit. The alluvial ore is also panned. The other categories of materials of higher sieve sizes are winnowed, separating concentrate from tailings.

Processing technologies initially used at these mines were gravity separation techniques using shaking tables, jigs and mechanized pans. However the equipment were either vandalized or reached the end of their useful life.

Plate 11.13: Workers Panning Wolframite Ore at Nyamuliro Wolfram Mine (Photos by I. Ntujju).

Cobalt

Cobalt production is on large scale utilizing the by-products stockpile of Kilembe Copper mine. At the copper mine which was operational from the 1950's to early 1980's, flotation was used to separate out the copper mineral. However with cobalt mineral which occured with the copper ore., The cobalt was stockpiled as tailings since market prices were then not good. The current cobalt plant utilizes bioleaching technology to extract cobalt metal from the stockpiled material and finally producing cobalt chips from an electrolytic circuit. The plant also applies flotation in its separations.It is this tailings stockpile that is now being processed using a bioleaching process. The cobalt is processed into ingots for export.

Limestone

Limestone utilization processes are in two categories; cement and lime production. There are two plants producing cement: One produces 2000 tons per day and another 900 tons per day. The process in the 2000 tons per day plant involves: crushing limestone(marble) at quarry site to minus ½ inch, then blending(limestone from different sites) and feeding together with other raw materials:,(fluorspar and clay)to the raw mill; firing in a kiln to make clinker is out in a rotary kiln,.After clinker cooling, inter grinding clinker and gypsum is carried out in the cement mill to make cement.

In the 900 tpd plant, limestone is hauled from the quarry and crushed at the plant. The raw materials (limestone, clay, kzaolin and iron ore) are milled together before being fired in a kiln to form cement mill which is later on interground with gypsum to make cement.With the advent of mixed cements sometimes natural pozzolans are added to make Portland pozolan cements.

The difference between the two plants is in raw material constituents and kiln firing fuels. The 2000 tpd plant utilizes coal (~ 170 tpd) for its rotary kiln firing, whereas the 900 tpd plant utilizes heavy furnace oil.

Lime Production

Lime production is by mechanized and artisanal production methods. Mechanized production facilities produce lime to the order of 8,000 to 11,000 tons per year, whereas artisanal production is to the tune of 3.2 to 3.6 tons per month (Jennifer Hinton *et al.*, 2006).

There is one mechanized operation with well built charcoal fired kilns, a jaw crusher, hydration unit and air separation units (cyclones) with a bagging unit. This operation also has a regrind mill for cyclone underflow. However other operations have well built kilns which use firewood but manually carry out the following: crushing limestone, hydration, sieving and bagging. In all these operations the kiln is loaded in such a way that fuel (charcoal or firewood) are interchangeably loaded.

Artisanal lime production operations differ from the above category of operations in the kiln design. Here tunnel kilns are used. These are tunnels/trenches dug into the ground and firewood and crushed limestone interchangeable load in to and fired.

Gypsum

Gypsum production is by artisanal operations which do not do any processing at the moment but simply select the high grade pieces from those of lower grade. Indeed the way the gypsum is extracted from the ground is such that narrow holes are dug haphazardly while extracting what they visually perceive as valuable gypsum. These rat-holes are actually a hazard waiting to happen as so many of those

Plate 11.14: Kilns at Muhokya Lime Works.

Plate 11.15: Hydration Unit.

Plate 11.16: An Opening to One of the Pits at Kibuku.

that were previously worked are uncovered and hard to see because of the surrounding knee-high grass. However in the past the miners used to wash the material on screens to rid it of any clayey material and earth.

Vermiculite

The operational vermiculite mine was taken over by the multi-national company Rio tinto. However previously the mining was being carried out selectively and the process involved screening the mined material to rid it of unwanted material. The

Plate 11.17: Mining Gypsum Inside One of the Pits (Photos by I. Ntujju).

Plate 11.18: Namekhara Vermiculite Mine.

screened material was then blended with material from different sites and thereafter exfoliated.

Processing of Phosphates

Phosphates are presently processed on small scale.phosohate ore is quarried and transported to a shed where it is crushed and ground in a ball mil.The ground

Plate 11.19: Vermiculite Recovery Plant Site, 2005.

ore is then passed through a low-intensity magnetic separator to remove bulk of the iron ore.This activity is repeated until most of the iron ore is removed when up to 39 per cent p205 concentrate is achieved. Typical flow sheet is shown in next page.

Direct-Application Phosphate Rock

From the above operation phosphate concentrate of high value is obtained and this has been found to be suitable for use as a direct fertilizer which has helped in increasing agricultural yields where soil fertility has been identified.

Ore Treatment Processing of Wolfram

Site

The present mill was sited on the ore body. This has restricted mining in some ways. Plans to move the mill never materialized due to financial considerations. This has resulted into tailings disposal becoming difficult.

A new site must eventually be chosen taking into consideration:

- ☆ Minimizing trucking distance from the mine.
- ☆ Proximity of water resources.
- ☆ Room for tailings disposal as over 99 per cent of the ore from mine is tailings.

A site on the slope west of the mill appears to be most suitable

Feeders

Ore from the mine is dumped onto a stationary grizzly passing – 300 mm (- 12 inch) material. Men with sledge hammers reduce any ore size at this stage.

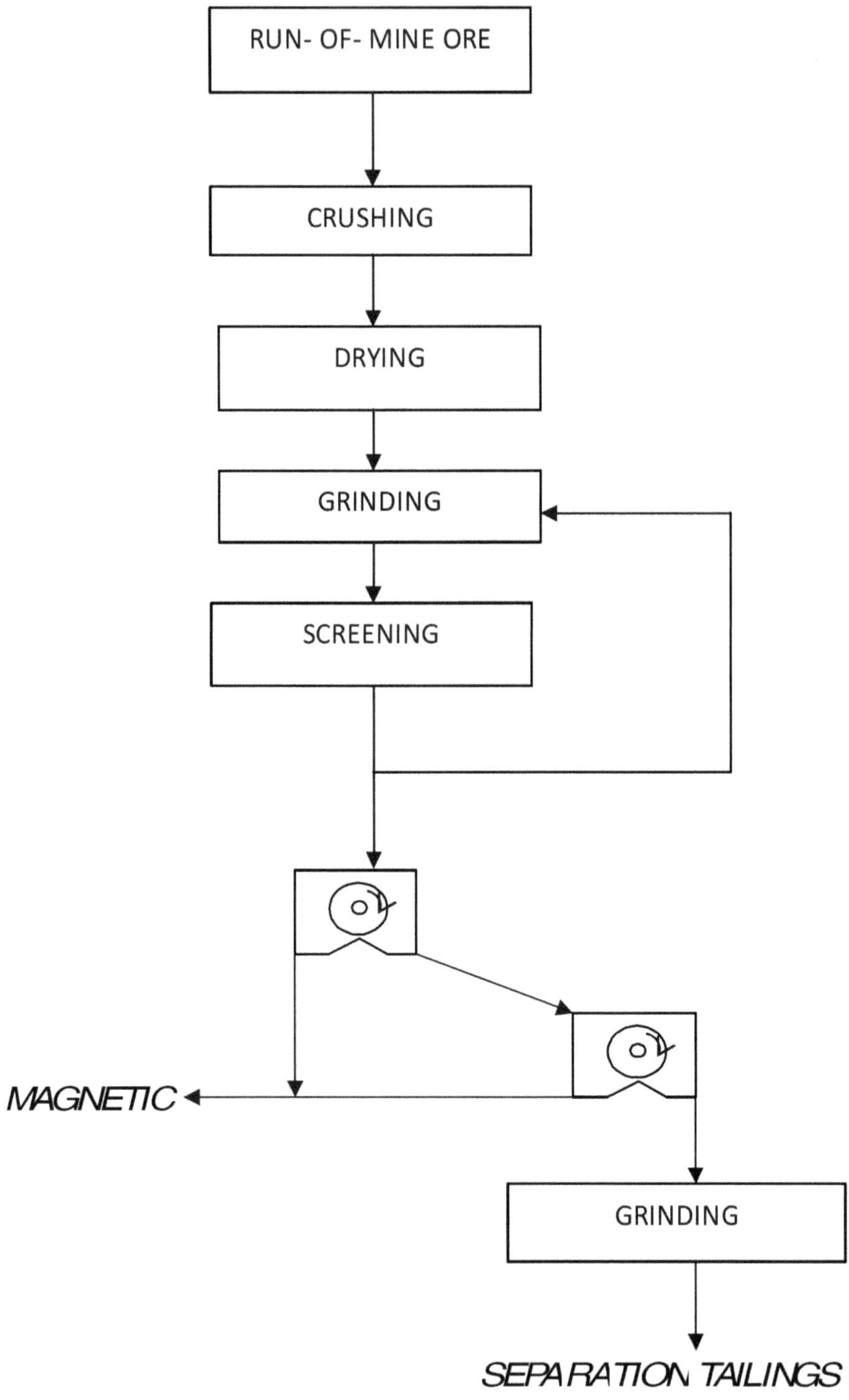
RUN- OF- MINE ORE
CRUSHING
DRYING
GRINDING
SCREENING
MAGNETIC
GRINDING
SEPARATION TAILINGS

Milling Flowsheet for Wolfram

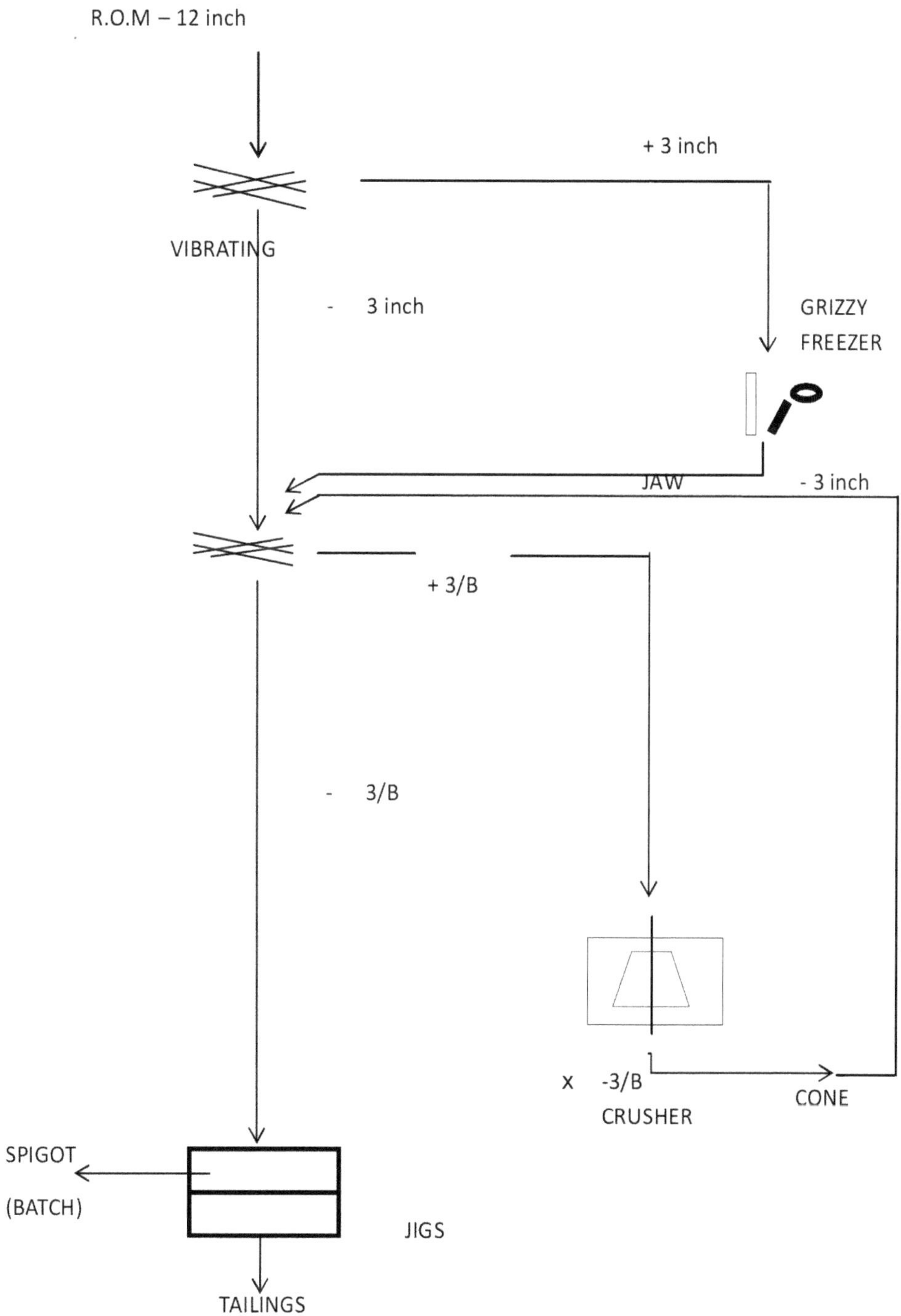

Crushing

For improved recovery the ore should be crushed finer than 20 mm (3/4 inch) as is being done now, to – 10mm (3/8 inch). Two stage crushing is necessary to crush - 10 mm (-3/8 inch). The jaw crusher would crush to -75 mm (3 inch) would be sized on a vibrating screen to -10mm (-3/8 inch). The =10mm (3/8 inch) would be crushed through a short head come crusher to – 10mm (3/8 inch). This material would be screened on the original deck hence forming a closed crushing circuit.

Concentration

The -10mm would be fed into two, 400 t.p.d Denver jigs designed to treat unclassified material.

The underflow -3.5mm (6 mesh or approx. 1/8 inch) would be treated in 2,200 t.p.d. secondary Denver jigs. The underflow fines approximately -2.5mm size (8 mesh) would be collected from the hutch of the jig as middling.

Tailing would be discarded at this jigging stage. Control at this stage is very crucial as over jigging would lead to loss of high grade gravels in the tailing and under jigging would result in low grade concentrates. The solution to this problem is not simple and experience appears to be the best answer.

The feed rate should be controllable to avoid chocking the jigs which would also result in losing wolfram with the tailings.

Results and Discussion

It is clear that Uganda has considerable mineral resources. The minerals are presently being processed to mainly concentrates which are shipped abroad without much value addition. The products of further upgrading from abroad are imported at high costs and jobs that would be created if upgrading was carried out locally are lost and lots of foreign currency that would have been gained lost. There is need for governments to establish policies that would enhance local skills in mineral processing technologies. The NAM countries could also enhance technical cooperation in these fields in a way of technology transfer. Nam countries can also encourage intra-nam investments in their respective mineral sectors.

Conclusions

There is great potential in the mineral sector of Nam countries. But to benefit from its potential there is need to invest in human capital to exploit this potential. A lot can be achieved if a mechanism was established for continuous exchange of ideas among experts in NAM countries and a scheme to encourage investments set up. It is for example possible to set up an ENam mineral Processing website for this kind of exchange. And this can start with this workshop.

References

1. Jennifer Hinton. 2005. A handbook of small scale mining in Uganda.
2. C. Lubangkene. 2006. Visit to gold ming Areas in Uganda.

3. Balu-Tabaaro. 1998. Francolitephosphate fertiliser project.
4. J.Arthur. 1977. Development of flowsheets for tin and wolfram.
5. F.K. Kallisaa. 1987. Development of industrial minerals in Uganda.

Chapter 12

Manufacture of Industry Spares from Copper Based Alloys: Development and Status in Zambia

C.V. Lungu

Department of Metallurgy and Mineral Processing, University of Zambia, PO BOX 32379 Lusaka, Zambia
E-mail: cvlungu@unza.zm

ABSTRACT

An opportunity exists to improve current efforts in the production of copper based alloys in Zambia through the study of mould filling characteristics and other factors associated with casting of these types of alloys. Copper based alloys are important in engineering applications besides the bulk iron and steel group and can deliver higher revenues to the manufacturer for comparable usages and quantities. Quality and reliability of castings is a direct result of microstructural development during the process of manufacture or casting.

According to Zambia's Commercial, Trade and Industrial Policy (January 2010) and published by the Ministry of Commerce, Trade and Industry, it highlights two major *policy objectives* for the engineering sub-sector and these are;

(i) To support the growth and expansion of the engineering products industry particularly into export markets and

(ii) To facilitate skills training and technological investments in the engineering products sub-sector.

The 2010 policy document further refers to three *policy prescriptions* to support the above objectives and critical among them is that the Government would support skills development by encouraging tertiary institutions to offer specialised training in metallurgy and related disciplines which are of relevance to the engineering sub-sector.

According to the report "*What would it take for Zambia's copper mining industry to achieve its potential?*" of July 2011 (UKAid and World Bank), improved linkages between the mines and local companies or suppliers can impact greatly on job creation as well as improving the quality of the jobs. The report further noted that with the copper industry output peaking at 1.5 million tonnes per year, Government annual revenue would increase to US$4.0 billion. The above report elaborated that Zambia's copper mines' cost of production was considered high due to the high cost of equipment and spares and other consumables.

Keywords: *Tin-Bronze, Copper, Casting, Mould-filling, Macro-segregation, Microstructure, Mechanical properties.*

Introduction

Zambia's Commercial, Trade and Industrial Policy (January 2010) and the Engineering Institution of Zambia (EIZ) Position Paper No. EIZ/2007/1 is among recent documents that spell out the country's ambitions and aspirations on wealth creation through value addition to locally produced raw materials. The Sixth National Development Plan (SNDP, 2011 to 2015) has targeted that the manufacturing sector is "*technology-based and export-focused manufacturing sector, which is dynamic and competitive with effective entities that add value to the locally abundant natural resources by 2030*". The goal of the manufacturing sector would be "*to develop a diversified and competitive export led value adding manufacturing sector which will contribute 12.5 percent to GDP*" (Summary document).

During the SNDP period, the strategic focus would be to strengthen and widen the country's manufacturing base with emphasis on backward and forward linkages given the country's wide resource base. The plan further elaborates that it would require intensifying the development of the resource-based industries, with the aim to optimise and add value to the country's natural resources.

One major SNDP objective, its strategies and programmes for the manufacturing sector are noted as shown in Table 12.1.

Table 12.1: The SNDP Objective and its Strategies

Objective	*Strategies*	*Programmes*
To expand the industrial base and increase value addition	(a) Facilitate the development of Multi-Facility Economic Zones/Industrial Parks; (b) Promote joint ventures between foreign and local investors; (c) Promote and facilitate Private Public Partnership (PPP) projects; and (d) Enhance the capacity of DRM.	Development of Multi-Facility Economic Zones and Industrial Parks

The establishment of Multi-Facility Economic Zones (MFEZ) in strategic locations in the country seeks to capitalise on easy availability of input raw materials for the manufacturing of a variety of consumer goods and industrial spares.

The EIZ Position Paper of 2007 (page vi) called for increased investment in manufacturing that targeted copper, gemstones, timber, agricultural commodities, iron ore and steel among others. There should be direct Government support in the establishment of metal powder plants, copper wire and rod mills as well as brass mills. The paper further stated that conventional financial support for such industries was unlikely to yield any good results and the best alternative therefore was for Government to consider loan guarantees and support locally bred initiatives.

Zambia as a country is positioned at the centre of the African craton or the metallogenic province which is dominated by Archaian to Neoproterozoic age rocks, which contain vast mineral resources (Nyambe, 2008). Nyambe further elaborated that the most important of these are the Katangan rocks of Neoproterozoic age which yield the copper and cobalt ores exploited on the Zambian Copperbelt accounting for over 80 per cent of Zambia's foreign exchange earnings. Coal bearing rocks of the Karoo (Permo-Carboniferous to Early Jurassic age) occur in the valleys such as mid-Zambezi Valley in the southern part of the country. The mining of gemstones and industrial minerals is also equally significant and the sector has the potential for enormous wealth and employment generation.

Production of Major Minerals and Metals

Ancient workings and those discovered in the 20th Century as well as the more recent ones at Lumwana in the North Western province and Konkola North on the Copperbelt (2005 to 2010) dominate the production of copper and other base metals of the non-ferrous variety. Iron ore and gemstones potential exist in many parts of the country as well and the Table 12.2 gives a summary of the mineral status for Zambia;

Table 12.2a: Known Major Mineral Resource Occurrences (After EIZ Position Paper, 2007)

Mineral Type	*Occurrence (Current and Future Exploitation)*
Cobalt	The entire Copperbelt and processed as a credit of copper production
Copper	The Copperbelt, Central, Luapula, Lusaka and North-Western Provinces
Iron	Nambala, Cheta and Sanje Hills, Namatombwa to the west of Lusaka
Lead and Zinc	Kabwe
Nickel	Munali Hills
Manganese	Luapula, Central and Northern Provinces
Gold	Central, Lusaka, Eastern, Luapula and Northern Provinces
Tin and Tantalum	Central, Northern, North Western and Southern Provinces
Uranium	Copperbelt, North Wesrtern and Southern Provinces
Phosphates	Nkombwa Hill, Chilembwe and Kaluwe
Amethysts, Emeralds and Aquamarine	Copperbelt, Eastern and Southern Provinces
Diamonds	Western Province
Coal	Southern Province in the Zambezi Valley

Table 12.2b: Major Producers, Country of Origin and Minerals Mined (Ministry of Mines, June 2012)

Name of Company	*Country of Origin (Shareholders)*	*Minerals Mined or Exploited*
Albidon Zambia	China and Australia	Nickel and minor copper
Chambishi Copper Smelter	China (Chinese govt.)	Copper
Chambishi Metals Plc	South Africa (Avmin)	Copper and minor cobalt
Chibuluma Mines Plc	South Africa (Metorex)	Copper and minor cobalt
First Quantum Minerals	Australia and Switzerland (Glencore)	Copper, minor cobalt and gold
Kagem Mining Limited	India/Zambia (Joint venture)	Emerald
Kariba Minerals Limited	Not verified	Amethyst
Konkola Copper Mines Plc	India (Vedantta)	Copper and minor cobalt
Konnoco	South Africa/Brazil (Teal Mining and Vale)	Copper and minor cobalt
Luanshya Copper Mines Plc	China (Chinese govt.)	Copper and minor cobalt
Lumwana Mines Plc	Canada (Barrick Gold)	Copper and minor uranium
Mopani Copper Mines Plc	Switzerland (Glencore)	Copper and minor cobalt
Non Ferrous China Africa Mining	China (Chinese govt.)	Copper and minor cobalt
Sable Zinc Limited	South Africa/Zambia (Metorex)	Copper
Tranter	Zambia	Manganese
Zamgem International Limited		Quartz

Note that the shareholding structure alters from time to time in response to international dynamics. There are numerous small scale miners producing copper and cobalt ores, industrial minerals and gemstones whose output generally ends up with the major companies and is accounted for as toll treated materials.

Table 12.2c: Annual Production of Major Metals and Minerals in Metric Tonnes

Metal or Mineral	*2010*	*2011*	*Comments*
Copper cathode	762,520	829,245	Includes figures from small scale producers
Cobalt flakes	1,328	1,208	Tonnage has declined sharply since the mid-1980s
Nickel concentrates	33,350	31,244 (kg)	Sole producer is mine at Munali Hills
Gold (toll treated)	1,985 (kg)	1,942 (kg)	Mainly by product gold
Emerald rough	–	35	Figures fluctuate a lot
Amethyst rough	2,155	19,545	Figures fluctuate a lot
Aquamarine rough	2,191	166 (kg)	Figures fluctuate a lot

Global demand for copper is known to outstrip supply which is expected to peak at 25 million tonnes in 2020 and that status is expected to maintain high metal prices till then at least (World Bank UKAid, 2011). If Zambia's current momentum in mining and mineral extraction investment is maintained and supported by a favourable regulatory framework, combined output of copper could reach 1,500,000 metric tonnes in 2020. The estimated annual export earnings would then reach around US$12 billion with revenues accruing to Government of US$4 billion per year.

It has been noted that during the year 2004 (*Export Board of Zambia, 2005*), the engineering sub-sector accounted for only US$65 million in export earnings whose major products included copper rods and cables, metal sheets, strips and coils. Zambia Metal Fabricators (ZAMEFA) accounted for 90 per cent of earnings in the sub-sector and the markets exported to were generally regional. The earnings from value addition to copper could substantially increase if specific steps are taken to support local production in the value chain.

While quality of the resources and worked deposits of the major metals remains impressive, mining input costs are considered very high as most large scale operations are in the upper quartile of cost curves. Labour productivity is also of particular concern just as is the input of mining equipment and consumables. Another area that is thought to be unstable is the issue of mining taxation and general regulation of the sector as that may work against new developments.

Mineral and Metal Led Industrialisation

Commercial exploitation of minerals in Zambia has taken place for over a century now and created what commentators describe as an "economic enclave" that is one that produces raw materials for other advanced economies while leaving no prospects for industrialisation of the mining host country. While growth and expansion of modern economies will continue to be led by good "know how" and rare skills, there is a lot of room for mineral led industrialisation and wealth creation. In the past decade an attempt was made on the Zambian Copperbelt to diversify from mineral production to agriculture in a Government led programme but little was achieved.

The Nature of Non-Ferrous Spares Used in the Mining Industry

Part of the motivation that led to the proposal on *"The effects of mould filling characteristics of 10 per cent tin bronze alloy on compositional segregation and metallurgical structure during sand casting"* was that the Zambian mining and allied industries are in great need for high quality and cost effective spares to service vital equipment. Many of the spares as alluded to in the introductory section are procured at great cost to mining operations and in some cases add costly lead times to production also. Local manufacturing of some identified spares in the semi-finished or finished form can improve the cost competitiveness of mining operations in Zambia.

Quality and reliability of castings is a direct result of microstructural development during the process of manufacture or casting. The issue of value addition to locally produced cathode copper has also gained national prominence during the start of 2012 and the University must lead in providing new knowledge in that direction.

The research topic therefore proposed to look at the factors that influence quality of castings in a relatively simple foundry environment is titled *"The effects of mould filling characteristics of 10 per cent tin bronze alloy on compositional segregation and metallurgical structure during sand casting"*. The major aim of the research is to model and test mould filling behaviour of a 10 per cent tin bronze using appropriate CFD techniques and determine the influence of mould filling characteristics on composition segregation as well as metallurgical structure on cast product quality.

One way of identifying the type of service components used in industry was to examine the testing schedule (or record) kept by a prominent analytical and testing laboratory, Alfred H Knight (AHK), situated on the Copperbelt where most mining and industrial activities are concentrated. AHK has operated in Zambia under that name since 2000 and prior to that the centre was called the Engineering Services section of the Zambia Consolidated Copper Mines (ZCCM) Group.

The components listed in Table 12.3 were a sample of the many that AHK has tested for various reasons dating as far back as 1980. Some of the testing requirements performed at AHK include; (i) localisation of component manufacture, (ii) failure analysis for warranty purposes, (iii) welding or joining procedure determination and (iv) quality assurance for procurement. All components contained at least 80 per cent copper with major alloying elements of tin, zinc, lead, aluminium and nickel.

Nearly all the items indicated in Table 12.3 were tested for the purpose of local manufacture needs in order to cut on long lead times associated with ordering spares from overseas. It was noted in the case of castings, a number of them were found to have failed prematurely for the reasons of porosity and second phase particles in the material structure that induced fatigue cracking. The sources of under-performing bronze based components referred to in Table 12.3 were from local suppliers in Zambia and outside the country.

Improved Mould Filling as a Way of Enhancing Casting Quality

One study on copper based alloys describes three examples of cast copper alloys and how technological properties and mechanical characteristics can be controlled by the metallurgical and casting processes (Stucky, 2007). Another study on "Mold filling and solidification of complex foundry castings" by Brown *et al.* (2002) is one example of studies primarily focusing on ferrous and aluminium alloy systems and established a direct linkage between quality of casting and mould filling behavior.

Solute segregation during solidification of a metal is mostly caused by the duration of the transition from liquid to solid (mushy zone) which is dictated by composition according to Baumeister *et al.* (2004) as the wider the transition is the greater the tendency for solute segregation.

Solute segregation also leads to changes in grain morphology as noted in secondary dendrite arm spacing and reported by Vander Voort (1984) and he distinguishes between two types of segregation; micro and macro-segregation. Vander Voort further observed that the former was more of a natural phenomenon during solidification while macro-segregation was dynamic under the same conditions and could be influenced by control measures.

There is a direct relationship between a good and successful or sound casting on one hand and effective mould filling on the other. That is so because complete delivery of the available molten metal to the casting cavity is critical in attaining a defect free

Table 12.3: Examples of Components or Parts Commonly Used in the Mining Industry (Alfred H Knight, August 2012)

Sl.No.	Component or Part Name	Application	Alloy Composition and Standard	Method of Manufacture
1.	Mid-span joint	Contacts for overhead power transmission lines	99.47% Cu, 10ppm S, 7.95ppm Ag. *BS1433 Grade C101*	Drawing
2.	Copper wires	Motor rewinding	Not indicated	Drawing
3.	Bronze bush	Slurry pumps	Copper-Zinc: 4.79% Sn, 5.42% Zn, 4.97% Pb, 0.02% P, 0.03% S, 0.28% Ni, 0.14% Fe, <0.01% others. *BS1400: Grade LG2*	Casting
4.	Wear out sleeve	Not indicated	Copper-Tin: 9.6% Sn, 0.09% Pb, 0.10% Zn, 0.01% Ni, 0.14% P. *BS1400: Grade CT1*	Casting
5.	Worm wheel	Locomotive stock	Copper-Tin: 11.0% Sn, 0.17% Pb, 0.3% P, others<0.01%. *BS1400: Grade PB2*	Casting
6.	Electrical contact	Winder stator reverser	99.9% Cu, others <0.01%. *BS1433 Grade C101*	Drawing
7.	Blow pipe	Removing unexploded explosives from mining drill holes	98.7% Cu, 0.04% P, 0.03% Zn, others <0.1%.	Extrusion
8.	Bronze bush	Mobile equipment motor	88.45% Cu, 7.89% Sn, 0.04% P, others <0.01%. *ASTM B438-73*	Sintering
9.	Bronze bearing/ bush	Ball and rod mills in grinding	83.2% Cu, 8.33% Sn, 2.50% Zn, 2.80% Pb, 0.01% Al, 0.08% Fe, 0.21% Ni, 0.12% P. *BS1400 grade LG4*	Casting

casting with acceptable process yield (Brown *et al.*, 2002). There is a performance gap between components designed for optimum load bearing abilities and those designed for castability. The two requirements must be met adequately.

Inadequate control of or poor mould filling will result among others in defects such as cavities in castings, trapping of gases, inclusions and improper dimensions according to Jain (2004). The said defects and others lead to high product rejection rates and uncompetitive products. Jain (2004, p. 376) mentions that commercial software packages such as castCAE are available as tools for simulating mould filling and solidification processes. Such packages would greatly assist investigation of the mould filling in the proposed study set up. The dynamics of mould filling and molten metal surface tension combine to determine delivery of the metal to the cavity for a successful casting, according to Campbell (2004).

Way Forward on the Current Situation in Zambia

It is obvious from the goals set in the Sixth National Development Plan (2011 to 2015) and other supporting national initiatives such as (i) The Commerce, Trade and Industry Policy Implementation Plan (2010 to 2014), (ii) The Micro, Small and Medium Enterprise Development Policy and (iii) The Engineering Institution of Zambia position paper on Value Addition (2007) that localisation of manufacture of various types of spares such as those mentioned in Table 3.0a can greatly improve industry competitiveness and national income.

Acknowledgements

I would like to acknowledge contributions from the institutions and individuals for the their valuable support in terms of information and data;

(i) Ministry of Commerce, Trade and Industry (MCTI), Lusaka
(ii) Ministry of Mines, Energy and Water Development (MMEWD), Lusaka
(iii) Alfred H Knight Zambia Limited (AHK), Kitwe
(iv) Engineering Institution of Zambia (EIZ)
(v) Mr. C C Simukonda (MCTI), Lusaka
(vi) Mr. A Sichivula (AHK), Kitwe
(vii) Mr. D T Chilumbu (MMEWD), Lusaka

References

1. *Value Adding: The key to employment and wealth creation*, Position Paper No. EIZ/ 2007/1, Engineering Institution of Zambia (2007).
2. *What would it take for Zambia's copper mining industry to achieve its potential?*, World Bank, UKAid (2011).
3. *Sixth National Development Plan 2011-2015 Executive Summary*, Government of the Republic of Zambia (2011).
4. *Commerce, Trade and Industry Policy Implementation Plan 2010-2014*, Ministry of Commerce, Trade and Industry (2010).

5. *The Micro, Small and Medium Enterprise Development Policy*, Ministry of Commerce, Trade and Industry (2010).
6. Nyambe, I A, Kawamya V M, 2007, '*Approaches to sustainable minerals development in Zambia*', Geological Society of Zambia.
7. Stucky, M. (2007), *Relationship between metallurgy, process and properties of cast copper alloys, in Copper: better properties for innovative products (ed J.-M. Welter)*, Wiley-VCH Verlag GmbH and Co. KGaA, Weinheim.
8. Brown, SGR, Spittle JA, James JD 2002, 'The mold filling and solidification of a complex foundry casting', *Journal of Metals (JOM-e)*, viewed 21st October 2011 http://www.tms.org/pubs/journals/jom/0201/brown/brown-0201.html
9. Vander Voort, GF 1984, *Metallography: Principles and Practice*, ASM International.
10. Baumeister, G, Haußelt, J, Rath, S, Ruprecht, R 2004, *Microcasting*, Institute for Materials Research, Forschungszentrum Karlsruhe, Germany.
11. Jain, PL 2004, *Principles of Foundry Technology*, Tata McGraw-Hill Education, New Delhi.
12. Campbell, J 2004, *Castings practice*, Butterworth-Heinemann, Oxford.

Chapter 13

Bioleaching in Mineral Processing and Beneficiation

Isha Sehgal

Research Assistant,
Centre for Science and Technology of the Non-Aligned and
Other Developing Countries (NAM S&T Centre), New Delhi, India
E-mail: sehgalisha305@gmail.com

ABSTRACT

Biotechnology plays an important role in mineral processing and beneficiation process. With the help of microorganisms, a variety of techniques are available and are being researched upon for the extraction of ores containing metals in minimal amount which otherwise may not be feasible with the traditional methods used for metal extraction. Bioleaching is a process that causes no pollution and is more environment friendly than the traditional extraction methods because it eliminates the steps like roasting and smelting of the traditional methods and also does not produce any offensive gases as it is a hydro metallurgical form of treatment. Besides this, the latter techniques are usually very expensive, energy intensive and environment unfriendly. Bioleaching is an inter-disciplinary field involving metallurgy, chemical engineering, microbiology and molecular biology. It has tremendous practical application and particularly for the developing countries, it has great significance where there are vast unexploited mineral resources.

The most commonly used microorganisms for bioleaching are *Thiobacillus thiooxidans* and *Thiobacillus ferrooxidans*, which are chemolithotrophic bacteria converting the insoluble metal sulphides into soluble metal sulphates. They usually act either by coming in contact with the ores containing minerals and causing oxidation of minerals or by producing leaching agents which oxidize the metals. Another important feature of these microorganisms is that they obtain energy from reduced sulphur composites.

Bioleaching can also be used for detoxification of municipal and industrial wastes containing toxic metals which can eventually diffuse into the underground water and pose serious threat to health. The thiobacillus strains grow well in waste water sludge and bring about its detoxification while recovering the valuable minerals. Therefore Bioleaching technology can prove to be very promising for all the nations worldwide, particularly the developing countries which are rich in mineral resources as they will have a competitive edge in this area contributing to economic and social development in the country.

Keywords: *Bioleaching, Roasting, Smelting, Mineral processing, Beneficiation, Extraction methods.*

Introduction

Bioleaching is not a new method of metal extraction, but a very old process which dates back to Roman times in the first century BC. The miners in those days used to leach out copper from its ore with the help of microbes but they were unaware of the microbial activity involved in the process. But now, with the research activities in this area, scientists are getting potential results which have brought about revolution in the mining industry.

The high grades world over are depleting at a fast pace due to the high demand for metal and the processes of roasting and smelting, the steps involved in metallurgy are not proving to be economical. Here, the microorganisms come into picture bearing a clear advantage over it. Microbes not only provide an economical method of extracting metals but they are also eco-friendly and do not cause pollution unlike the traditional methods of metal extraction (Mohd. Haris Siddiqui, Ashish Kumar, Kavindra Kumar Kesari and Jamal M. Arif, 2009).

In recent years, it has also been discovered that microbes find an application in the treatment of municipal and industrial wastes while recovering some minerals from the wastes. They aid in waste management by detoxifying the materials for safe deposition in the environment (Bajirova and Velkova, 2001; Bosecker, 2001; Soliso *et al.*, 2002; Picher *et al.*, 2002; Aung and Ting, 2005; Ishigaki *et al.*, 2005; Santhiya and Ting, 2005).

Bioleaching is a biological process in which there is an interaction between the microbes and the metal ores. As a result of this interaction, the metals are dissolved from their ores and recovered. These metals are dissolved through transformation by oxidation and reduction of organic and inorganic acids with the release of complex making agents (Barrett *et al.*, 1993; Bosecker, 1997; Sand *et al.*, 2001; Tribustch, 2001).

As mentioned earlier, metals are solubilised by means of oxidation and reduction reactions. Oxidation reactions are those reactions where a substance loses its electrons with the element part of ore getting separated into precious metals and the unused sulphur and acidic chemicals(Ref: http://www.innovateus.net/earth-matters/what-bioleaching) whereas reduction reactions are those reactions where a substance gains electrons and gets reduced.

The microorganisms involved in the process are iron- and sulphur-oxidizing bacteria like *Acidithiobacillus ferroxidans*, the sulphur-oxidizing *Acidithiobacillus thiooxidans* and *Acidithiobacillus caldus* and the iron-oxidizing *Leptospirillum ferroxidans* and *Leptospirillum ferriphilum*. Other than these bacteria, some species of fungi are also involved in bioleaching. They are *Aspergillus niger* and *Penicillium simplicissimum* which can mobilize copper and strontium by 65 per cent whereas aluminium, nickel, lead and zinc by more than 95 per cent (Brauer, H., 1991).

Bacterial Bioleaching Mechanism

The process of bioleaching involves either of the two types of methods to carry out the solubilisation of metals from their ores. They are the following methods:

1. Direct Bioleaching
2. Indirect Bioleaching

1. Direct Bioleaching

In this type of bioleaching, the microorganisms have a direct contact with the surface of mineral ore and the direct attachment occurs by the establishment of close physical contact between them resulting in sulphates through a series of enzymatically catalyzed reaction steps (Barrett *et al.*, 1993; Bosecker, 1997).

For example; $4FeS_2 + 15O_2 + 4H_2O \xrightarrow{\text{Bacteria}} 2Fe_2(SO_4)_3 + 2H_2SO_4$

According to Bosecker, 1997; Sand *et al.*, 2001; Tribustch, 2001, bacteria does not attach to the whole mineral but only a specific area.

2. Indirect Bioleaching

In this type of bioleaching, there is no direct contact between the microorganism and the mineral ore, instead the microbes produce metabolites like organic and inorganic acids which act as a leaching agent and leach the mineral.

It involves the culturing of microorganisms in an appropriate culture medium, giving them optimum conditions of low pH and high temperature to allow for the production of metabolites in adequate quantity which is later used to leach out the mineral from its ore.

An example of indirect bioleaching is given as under:

$$FeS_2 + Fe_2(SO_4) \longrightarrow 3FeSO_4 + 2S^o$$

$$2S^o + 3O_2 + 2H_2O \longrightarrow 2H_2SO_4$$

Fungal Leaching Mechanism

Fungal leaching involves three different mechanisms as follows:

1. Acidolysis: It is a process by which the fungi cause dissolution of minerals by excreting protons and organic acids which in turn attack the mineral surface.
2. Complexolysis: In this process, solubilisation of metal occurs by the formation of metal complexes by the fungi which increase the metal mobility.
3. Redoxolysis: In this process, metals are oxidized or reduced by oxidation-reduction reactions or by complex formation leading to increased metal mobility thereby causing its solubilisation.

Microorganisms in Bioleaching

There are three main groups of microorganisms involved in the process of bioleaching. They are autotrophic bacteria, heterotrophic bacteria and molds. The following examples of microorganisms will give a clear picture of their role in the bioleaching process:

Thiobacillus

- They are gram negative, non-spore forming, rod shaped aerobic bacteria, Figure 13.1.
- Their cell measure about 0.3 and 0.8 µm in diameter and between 0.9 and 2 µm in length.
- They are capable of oxidation of sulphur, most of them are chemolithotrophic as they use CO_2 as an energy source.
- Leaching is carried out in the range of 1.5-3 pH *i.e.* they can tolerate high acid levels.
- Important strains are *Thiobacillus ferroxidans* and *Thiobacillus thiooxidans* (Barrett *et al.*, 1993).
- They obtain energy from oxidation of reduced sulphur composites.
- *T.ferroxidans* is very similar to *T.thiooxidans* morphologically but differs from *T.thiooxidans* in slow oxidation of elemental sulphur.
- Apart from these, *Acidophilic thiobacillus* strains were also discovered.

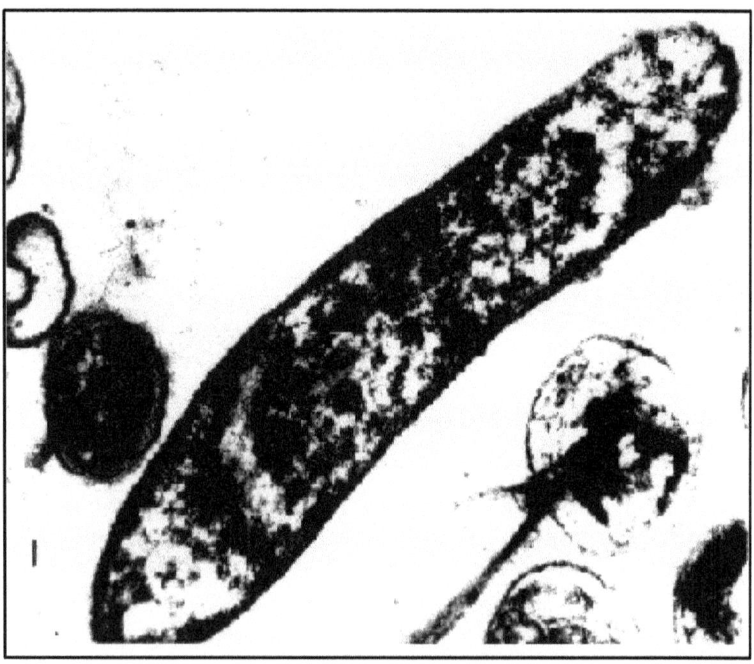

Figure 13.1: A Photograph of Thiobacillus.
***Source*: http://microbewiki.kenyon.edu.**

Leptospirillum

- ☆ They are aerobic microorganisms with polar flagellums belonging to Spirillaceae famil, Figure 13.2.
- ☆ *Leptospirillum ferroxidans* is acidophilic and obligatory chemolithotrophic bacteria that can oxidize iron.
- ☆ They can tolerate low pH but are unable to oxidize sulphur and sulphur composites.
- ☆ They are very sensitive to copper and can tolerate high concentrations of uranium, molybdenum and silver (Barrett *et al.*, 1993; Leduc and Ferroni, 1994; Krebs *et al.*, 1997).

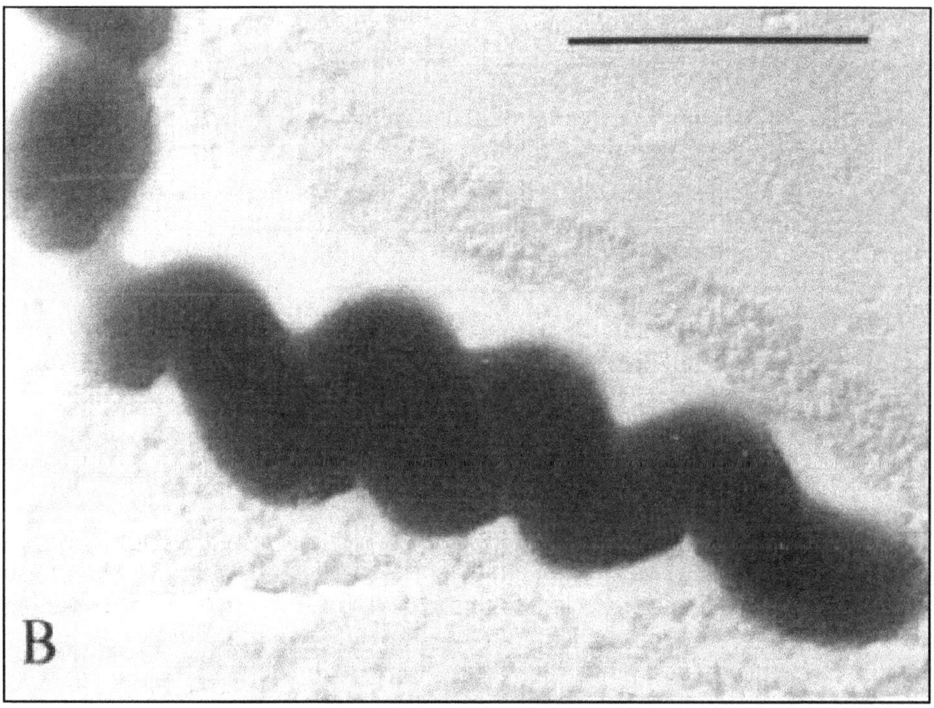

Figure 13.2: Leptospirllum.
***Source*: http://microbewiki.kenyon.edu.**

Thermophilic Bacteria

- ☆ They are aerobic, facultative and chemolithotrophic bacteria, Figure 13.3.
- ☆ These can grow at 50°C on metals and use ferrous ion as a source of energy.
- ☆ *e.g. Acidionous bricleyi* is a chemolithotrophic, facultative aerobe.

Heterotrophic Microroganisms

- ☆ These microbes require organic carbon and energy to carry out their functions.

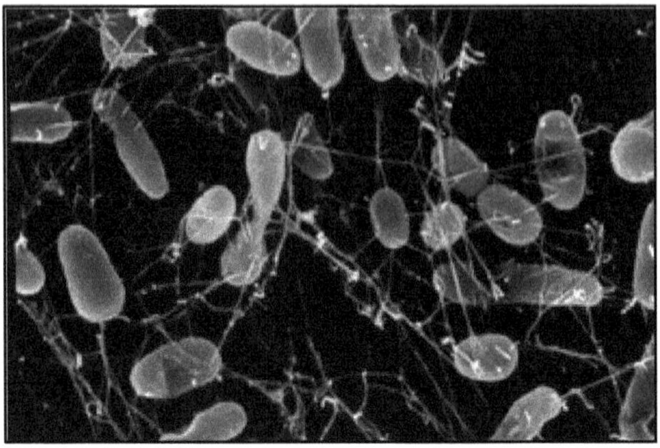

Figure 13.3: A Photograph of Thermophilic Bacteria.
***Source*: http://www.sciencecentric.com.**

- They are mostly filamentous fungi and bacteria which act by producing organic acids, amino acids as well as other metabolites which brings about the dissolution of metals from their minerals.
- This type of bioleaching is favourable for minerals which are low in sulphur and sulphides with low pH value.

Commonly Used Microoganisms

Some of the microorganisms which are commonly used in bioleaching are:

- *T.ferrooxidans* and *T. thiooxidans*: *Thiobacillus ferrooxidans* is a gram negative, rod-shaped, motile, non-spore forming bacterium. It derives the energy for the growth from the oxidation of iron or sulphur in ore. This microbe oxidises ferrous ions (Fe^{+2}) to ferric ions (Fe^{+3}) and converts sulphides, thiosulphate, sulphur to sulphate. *Thiobacillus thiooxidans* has similar functions and it grows mostly on sulphur ores or rocks.
- *Sulfolobus acidocaldarius* and *Sulfolobus brierlevi* are involved in bioleaching of copper and molybdenum from ores like chalcopyrite and molybdenite. They are thermophilic and acidophilic bacteria which mostly grow in acidic hot springs.
- Bacteria like *Pseudomonas aeruginosa* are used to extract high value metal like uranium from low quality uranium ore containing around 0.02 per cent uranium. This microorganism extracts around 100 mg of uranium per one litre solution in less than 10 seconds.
- *Rhizopus arrhizus* is another bacterium that can extract uranium from low grade uranium ores.

Some of the fungi which are associated with bioleaching are *Aspergillus niger*, involved in leaching of copper and *Aspergillus oryzae*, carry out the extraction of gold.

Commercial Extraction of Minerals by Bioleaching

The commercial extraction of metals by bioleaching involves the process to be optimized by controlling factors like temperature, pH, humidity, oxygen and carbon dioxide concentration. The microorganisms along with the required nutrients and acids are dumped into the ore bed thereby leading to the extraction of leach liquor which is then collected for the recovery of metals.

There are three types of processes of commercial extraction of metals, which are as follows:

1. Slope Bioleaching

In slope leaching the ore is finely ground and kept in large pile in a slope which is subjected to continuous sprinkling of aqueous solution of microorganisms.

The leach liquor collected at the bottom of the ore is processed further for metal recovery.

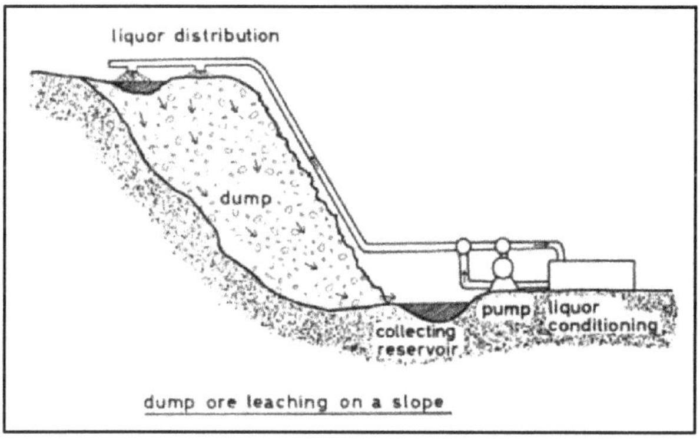

Figure 13.4: Shcematic Diagram of Slope Bioleaching.
***Source*: R. Näveke, Institut für Mikrobiologie,**
Technische Universität Braunschweig, Fed. Rep. Germany.

2. *In-situ* Bioleaching

Aqueous solution of microorganism is pumped through drilled passages within the ore. The leach liquor collected at the bottom of the ore used for metal extraction.

3. Heap Bioleaching

In heap leaching the ore is arranged in heap and goes through the same treatment such as in slope leaching. The aqueous solution containing microorganisms works on the heap of ore and produces the leach liquor. The leach liquor is used for metal recovery.

Recovery of Metals from different Types of Waste Materials

It is possible to recover metals from different types of waste materials like industrial and municipal wastes. In this process, the heavy metals are mobilized by

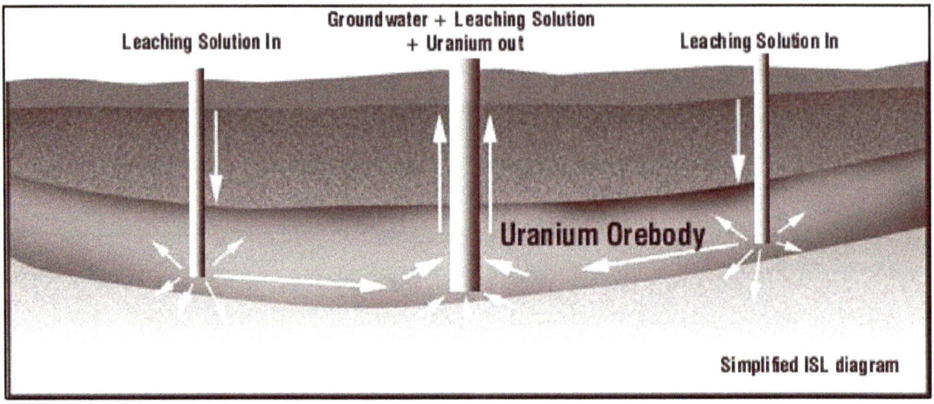

Figure 13.5: A Chemicatic Diagram of *In-situ* Bioleacing.
***Source*: www.google.co.in.**

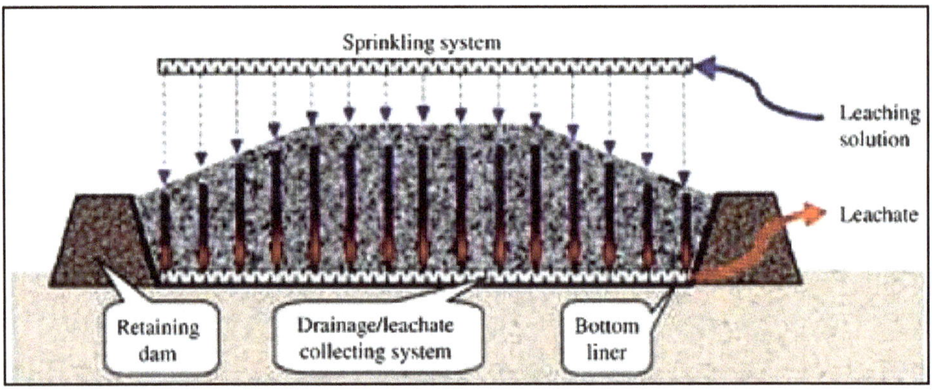

Figure 13.6: A Schematic Diagram of Bioleaching.
***Source*: http://www.eurssem.eu/pages/4-5-9-ex-sito-technologies-srt.**

microorganisms and separated leading to detoxification of waste materials. Hence, it is now safe for disposal into the environment (Bosecker, 2001).

A Table 13.1 shows the role of different types of microorganisms for recovery of metals from different waste types.

Advantages and Disadvantages

There are many advantages and few disadvantages of bioleaching as follows:

Advantages of Bioleaching

1. Bioleaching allows for the extraction of metals from even low grade ores which do not contain high concentration of metals.
2. It is a very economical process as the operating cost is quite low unlike the traditional methods which require the highly expensive processes of smelting and roasting.

Table 13.1: Different Types of Microorganisms for Recovery of different Metals

Waste Type	*Rocovered Metals*	*Microorganisms*	*References*
Sewer sludge	Heavy Metals	*Thiobacillus* sp.	Bosecker. 2001; Pidier *et al.*, 2002
Filter dusts	Cu, Cr, Zn	*Thiobacillus thiooxidans*	Krebs *et al.*, 1997
Fly ash from municipal waste	Zn, Ni, Cu, Zn, Cd, Cr, As	*Thiobacillus thiooxidans, Aspergillus niger*	Ishigaki *et al.*, 2005
Pig manure	Cu, Fe, Zn	*Acidothiobacillus ferrooxidans*	Picher *et al.*, 2002
Mine waste	Al, Na, Mg, K, Zn	*Thiobacillus ferrooxidans*	Bojinova ve Velkova, 2001; Solisio *et al.*, 2002
Paper factory waste	Fe, Cu, Zn	*Acidothiobacillus ferrooxidans*	Picher et al., 2002
Petroleum refinery waste	Al, Mo, Ni	*Aspergillus niger*	Aung ve Ting, 2005; Santhiya ve Ting, 2005

Source: Hacettepe Journal of Biology and Chemistry, Volume 35, Issue 1, January 2007, pp. 17-24.

3. It can also be employed to extract metals from wastes or drainages.
4. It is an environment friendly method as it does not produce any harmful gases.
5. It can also be used to extract a refined and expensive metal which is not possible by other chemical processes.

Diadvantages of Bioleaching

1. It is a very slow process as compared to the traditional methods and requires lot of time because of the time taken by the microorganisms for their activity.
2. As bioleaching is dependent on several factors like atmospheric conditions, it decreases to an extent the efficiency of the process.

References

1. Aung, K.M. and Ting, Y. P. (2005). Bioleaching of spent fluid catalytic cracking catalyst using Aspergillus niger. *Journal of Biotechnology*. 116, 159-170.
2. Barrett, J., Hughes, M. N., Karavaiko, G.I. And Spencer, P.A. (1993). Metal extraction by bacterial oxidation of minerals. ss. 1-153, Ellis harwood limited, England.
3. Bojinova, D. Y., And Velkova, R. G. (2001). Bioleaching ofmetals from mineral waste product. *Acta Biotechnol*. 21, 275-282.
4. Bosecker, K. (1997). Bioleaching; solubilization by microorganisms. *FEMS Microbiology Reviews*. 20, 591-604.
5. Bosecker, K. (2001). Microbial leaching in environmental cleanup programmes. *Hydrometallurgy*. 59, 245-248.

6. *Hacettepe Journal of Biology and Chemistry*, Volume 35, Issue 1, January 2007, pp. 17-24.
7. Ishigaki, T., Nakanishi, A., Tateda, M., Ike, M. and Fujita, M. (2005). Bioleaching of metal from municipal waste incineration fly ash using a mixed culture of sulfur-oxidizingand iron-oxidizing bacteria. *Chemosphere*. (In press)
8. Krebs, W., Brombacher, C., Bosshard, P. P., Bachofen, R. and BrandI, H. (1997). Microbial recovery of metals from solids. *FEMS Microbiology Reviews*. 20, 605-617.
9. Leduc, L. G. and Ferroni, G. D. (1994). The chemolithotrophic bacterium *Thiobacillus ferrooxidans*. *FEMS Microbiology Reviews*. 14, 103-120.
10. Mohd. Haris Siddiqui, Ashish Kumar, Kavindra Kumar Kesari and Jamal M. Arif. Biomining- A Useful Approach Toward Metal Extraction. *American-Eurasian Journal of Agronomy* 2 (2): 84-88, 2009.
11. Picher, S., Drogui, P., Guay, R. and Blais J.F. (2002). Wastewater sludge and pig manure used as culture media for bioleaching of metal sulphides. *Hydrometallurgy*. 65,177-186.
12. R. Näveke, Institut für Mikrobiologie, Technische Universität Braunschweig, Fed. Rep. Germany
13. Sand W., Gehrke T., Jozsa P. And Schippers A. (2001).Biochemistry of bacterial leaching-direct vs. indirect bioleaching. *Hydrometallurgy*. 59, 159-175.
14. Santhiya, D. and Ting, Yen-Peng (2005). Bioleaching of spent refinery processing catalyst using *Aspergillus niger* with high-yield oxalic acid. *Journal of Biotechnology*. 116, 171-184.
15. Solisio, C., Lodi, A. and Veglio F. (2002). Bioleaching of zinc and aluminium from industrial waste sludges by means of *Thiobacillus ferrooxidans*. *Waste Management*. 22, 667-675.
16. Tribustch, H. (2001). Direct versus indirect bioleaching. *Hydrometallurgy*. 59, 175-185.
17. http://csp.eworlding.com/3r/congress/manu_pdf/129.pdf
18. http://microbewiki.kenyon.edu
19. http://www.eurssem.eu/pages/4-5-9-ex-sito-technologies-srt
20. http://prr.hec.gov.pk/Chapters/876S-2.pdf

Part II

Policies and Strategies

Chapter 14
The Mineral Industry of Egypt

Ahmed Yehia

Central Metallurgical R&D Institute,
P.O.Box 87, Helwan, Cairo, Egypt
E-mail: yehia522000@yahoo.com

ABSTRACT

Maximum utilization of natural resources for different industrial applications is becoming a key issue in all countries for sustainable development and a balanced sheet of payment. The Egyptian mineral industry has enormous growth potential for the next decade owing to the country's abundant mineral resources- especially gold, natural gas, phosphate rock, and tantalum and the increased demand for steel reinforcing bar and cement by the construction sector. Egypt is rich with different mineral commodities but with a modest quality. Improving the technical specifications of such natural resources will not only result in rationalization of such resources but also, in improving the quality of the end products and enhancing their competitiveness in the world market. As for investment, Egypt possesses many investment attraction factors. These factors include; a strong infrastructure of transportation and communication, cheap energy sources, skilled manpower, eligible industrial cities, a strong and secure banking system, and an active stock market. Therefore, many experts consider Egypt's economy to be in a competitive position among the advanced world economies.

Keywords: *Mineral, Economy, Gold, Iron, Ilmenite, Sulphur, Egypt.*

Introduction

The Egyptian mineral industry has enormous growth potential for the next decade owing to the country's abundant mineral resources especially gold, natural gas, phosphate rock, and tantalum and the increased demand for steel reinforcing bar and cement by the construction sector [Taib, 2010]. Domestic demand for cement and steel is expected to increase in the short term as 524,000 low-cost residential units are being built throughout the country and expected to be completed by 2012.

Egypt expects to be a major producer of gold and tantalum in the next 5 years. The passage by the newly elected Parliament of the proposed mining law, which was drafted with the help of the World Bank, could reduce bureaucracy, enhance transparency, and facilitate financing to mineral industry projects. The production volumes of cement, phosphate rock, phosphate and nitrogen fertilizers and steel, are expected to increase to satisfy domestic demand for these mineral commodities.

Economy in Egypt

The Egyptian economy is an outcome of successive stages that targeted comprehensive structural economic reforms, especially during the last 25 years, to move towards a market -oriented economy, to liberalize stock market and to encourage the private sector which contributes by more than 70 per cent in the Gross Domestic Product [World FACT Book, 2012].

In the domain of economic foreign relations, Egypt moves towards trade liberalization through its membership in World Trade Organization, and participating in many regional and international economic groupings like the Egypt-EU Partnership Agreement, the Agreement to Facilitate and Develop Inter-Arab Trade with a view to establish the Arab Common Market, the COMESA group, the African Union, the Agadir Agreement, the G-8 Islamic countries, and the G-15 developing countries, in addition to many bilateral free trade agreements.

As for investment, Egypt possesses many investment-attracting factors. These factors are: a strong infrastructure of transportation and communication, cheap energy sources, skilled manpower, eligible industrial cities, a strong and secure banking system, and an active stock market, in addition to a climate of political, legislation and economic stability.

Therefore, many experts consider Egypt's economy to be in a competitive position among the advanced world economies.

Economic Reform

This stage started since 1990 up till now in the framework of changing to mechanisms of the market and adopting many policies to treat structural and monetary imbalances, the most important of which are: liberation of bank interest rates and exchange rates, establishing a free exchange market, implementing privatization programmes and liberalizing public sector and foreign trade.

In the 1990's Egypt exerted great effort to change from directed economy to market-oriented economy which contributed to a drop in the deficit and a reduction in the inflation rate to less than 3 per cent and in stabilizing exchange rate, in addition to trade liberalization, eliminating investment obstacles resulting in the increase of growth rate to 5 per cent.

Income Tax

Tax law no.91 of 2005 reflects a new thought in the tax system's dealing with tax payers.The new Tax Law represents a qualitative leap and a new turning point in the Egyptian national policy.This law contributes to reducing the tax categories to about

50 per cent or less for the benefit of the citizens and those working in economic activities. It provides many privileges, the most prominent of which are: tax exemption, tax conciliation, eradicate tax complications together with building bridges of mutual understanding between the state and tax payers in a manner that encourages local and foreign investment in Egypt.

Corporate Tax

Corporate taxes law targeted the attraction of all kinds of investments through a package of tax exemption and through postponing taxes for periods of time that ranges from 5 to 20 years according to the geographic location, after which the income is subjected to taxes. In agreement with the global approach to attract domestic and international investments, the corporate tax was comprehensively improved in order to:

- Facilitate investment and production.
- To encourage technology conveyance and reducing its cost.
- To provide continual resources of public revenues.
- Activate the ideals of tax fairness.

Minerals in the National Economy

In 2010, the Egyptian economic activity of the mining sector was 13.7 per cent of the country's gross domestic product (GDP). The industrial sector, which included fertilizer manufacturing and petroleum refining, accounted for 15.4 per cent of the GDP. Construction and building activity accounted for 5.2 per cent of the GDP [Central Bank of Egypt, 2010].

The flow of foreign direct investment (FDI) played an important role in Egypt's mineral industry which received about 69 per cent and 75 per cent of the total FDI in 2010 and 2009, respectively. The outflow of FDI from Egypt to the rest of the world increased by 106 per cent to $1,176 million from $571 million in 2009 [Arab Investment, 2011].

National Mineral Policy and Future Outlook

Over the past century, technology advances have had major impacts on mining practices and the nature of the mine in the world. It is known that productivity gains tend to increase as value is added to a product while it moves down stream. However, as processing plants become more highly tuned and as the just-in-time delivery principle becomes more important, the quality and stock of feed materials become more important and stock of feed material become more important determinants of plant performance. This suggests that R&D and innovations targeted at beneficiation processes are likely to have higher payoffs in the future and thus merit closer attention. The need for collaboration between technology suppliers and operating companies is clear. Advanced technologies are increasingly complex and R&D needs are often too great for a single company to support. In short, R&D roles, as well as funding levels, need to be reappraised in the industry to meet the technology objectives viewed as important by the industry itself. The National Mineral Policy would put Egypt in competition with other nations that also encourage mining investment. The resources are there. All that is needed to return Egypt to prominence in mining is increased government support for investment, exploration and development in the mining sector. Besides, in the mean time industry is playing a main role on solving the problem of unemployment by creating new jobs, particularly in the newly founded cities, *e.g.* 6 October city and tenth of Ramadan city. Therefore, for promoting the momentum of development in this industry and increasing the exportation capacity in this field, the following recommendations should be taken into consideration:

1. Means to guarantee the consistency and homogeneity of the raw materials quality and international code compliance.
2. An economically/politically healthy climate to encourage foreign investors, *e.g.* less tax, exemption from customs, subsidizing infrastructures and barging facilities, long term land tenure, stability taxation etc.

Geology of Egypt

The Egyptian territory is covered by crystalline basement rocks belonging to Precambrian age and Phanerozoic sediments which range in age from Cambrian to Recent (Figure 14.1). The basement rocks form about 10 per cent of the land surface and are exposed in South Sinai, Eastern Desert, and South West corner of Egypt [Said, 1990].

The basement rocks of Southwest corner of Egypt crop out from Egyptian-Libyan borders to Gabal Kamel as continuous low land and ridges, [Richter, 1986; Richter,1990]. From Gabal Kamel to Aswan, the basement occurs as uplifting inliers. The main economic minerals in this region are Banded Iron Formation probably of Lake Superior Type [Abu Salem, 2003Khattab, 2002; Abdel Fattah, 2005].

Mesozoic sediments are very unequally distributed. Marine Triassic is found in Aref El Naga, whereas, continental covers many areas in Egypt. The Jurassic age is well developed in Gabal Maghara and South West of Sinai, Khashm El Galala. Cretaceous sediments are widely distributed and form about 40 per cent of the Egyptian surface. The deposition of Cretaceous sediments is not only governed by regression and transgression of fall and rise of sea level, but also by renewal uplift of source areas and variations in continuous input linked to tectonism along the continental margin [Sloss, 1979] and within the craton [Van Houten, 1980]. All phosphate deposits and white sands are the economic minerals in Mesozoic.

Minerals Commodities in Egypt

The mineral commodities can be classified as metallic and non-metallic deposits [Khalid, 2002]. The most important of these deposits are:

1. Metallic ores such as: iron ores, gold ores, industrial metal oxides (Sn, Ta, No, W, and Mo), titanium and titaniferous-iron ores, manganese ores, sulphide mineralization (Pb, Zn, Cu, and Co), and chromite.
2. Non-metallic ores such as phosphate, limestone, dolomite, ornamental stones, quartz rock, white sands, talc, feldspar, kaolin, fire clay, bentonite, gypsum, fluorspar, sands and gravel, magnesite, evaporates (salts) and coal.

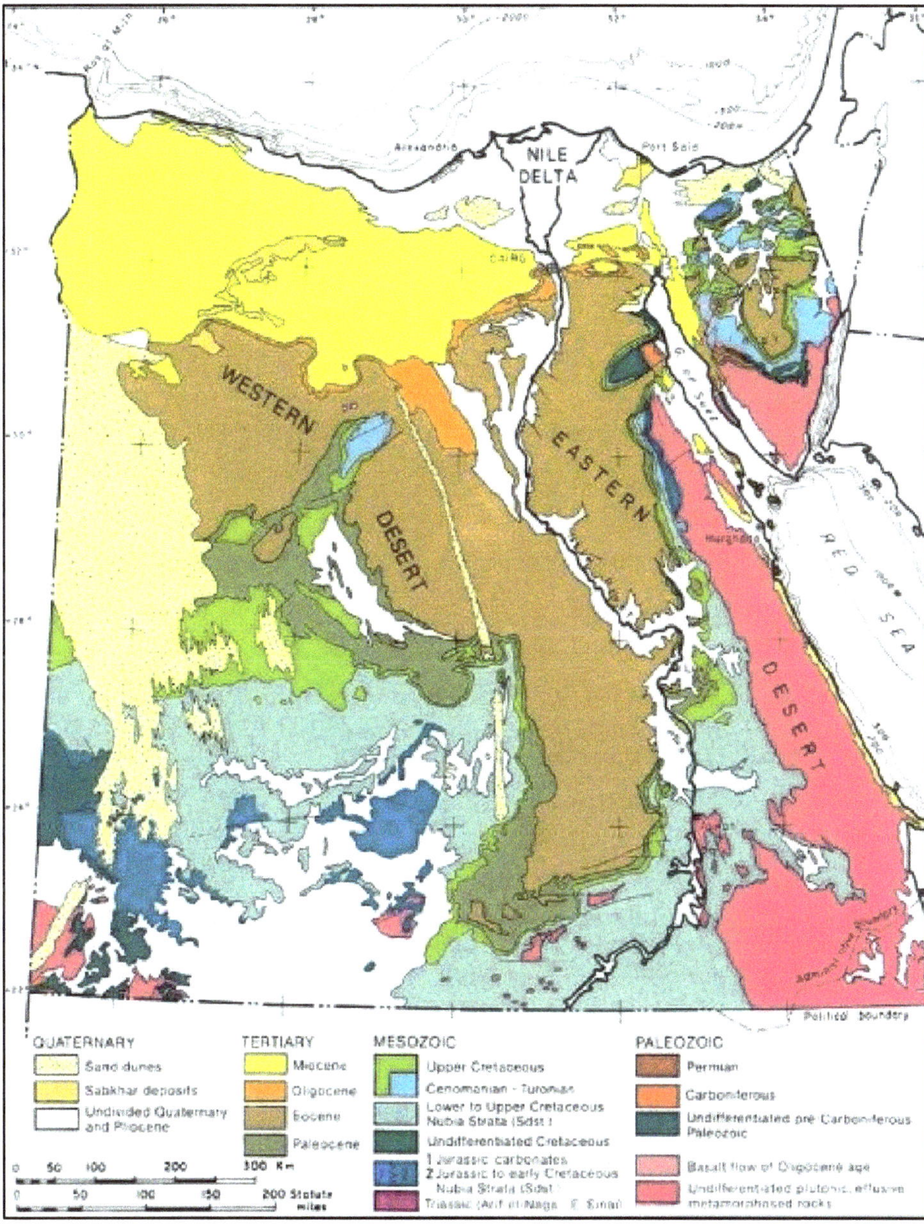

Figure 14.1: Geological Map of Egypt [Said, 1990].

Phosphate Rock

Egypt's major industrial mineral export is phosphate. In 2010, Egypt's output of phosphate rock decreased by about 26 per cent to 4.6 Mt from 6.2 Mt in 2009. El Nasr Mining was the main producer of phosphate rock in Egypt. The company, which had the capacity to produce 3 Mt/yr of phosphate rock, produced 2.9 Mt and exported about 2.3 Mt of phosphate rock from the East Sebaya Mine, the West Sebaya Mine, the Red Sea Mine at El Qusier, and the Abu Tartur Mine in the Western Desert. Other companies that produced phosphate rock included El Wataneya, which had the capacity to produce 2.1 Mt/yr, and Misr Phosphate Co., which had the capacity to produce 1.2 Mt/yr. In 2010, El Wataneya produced 252,000 t and exported 67,000 t, and Misr Phosphate Co. produced 271,000 t and exported 64,000 t [Arab Fertilizers Association, 2011]. Phosphate rock was transported by railway to Safaga Port on the Red Sea. El Nasr Mining owned two export ports on the Red Sea: Hamrawein Port, which had the capacity to load vessels of up to 35,000 t, and Abu Ghusun Port, which had the capacity to load vessels of up to 12,000 t. The New Valley phosphate project was one of the projects under development. It is located on the Abu Tartur plateau, at equal distance (650 km) from Cairo and the Port of Safaga on the Red Sea. Geologic explorations by EMRA indicated the presence of about 715 Mt of proven fresh phosphate rock reserves under an overburden 150 m thick, on average, that could be exploited underground using a fully mechanized long wall technique, as well as 20 Mt of weathered phosphate reserves that could be exploited using an open cast technique. Current production of phosphate rock at the Abu Tartur project was 1 Mt/yr of weathered phosphate [Arab Fertilizers Association, 2010].

Gold

In 2010, gold was produced from two mines in Egypt the Hamash gold mine and the Sukari gold mine. In January, the Sukari Gold Mine Co. began exporting gold to an overseas gold refinery. The company mined 6.2 Mt of ore, processed 3.1 Mt of ore, and produced 4,674 kilograms (kg) of gold (reported as 150,289 troy ounces). The Sukari Gold Mine was an open pit mine located about 23 km southwest of Marsa Alam in Egypt's eastern desert and was the first modern large-scale open pit gold mine operating in Egypt. The mineral resources of the mine as of yearend 2010 were estimated to include measured reserves of 84 Mt of ore grading 1.42 grams per metric ton (g/t) gold; indicated reserves of 151.7 Mt of ore grading 1.47 g/t gold; and inferred reserves of 68.9 Mt of ore grading 1.60 g/t gold. Probable and proven mineral resources of the Sukari Mine increased to 9.1 million troy ounces [283 metric tons (t)] of gold from 7.1 million troy ounces (221 t) of gold at a cutoff rate of 0.4 g/t for oxide and sulfide material and 0.5 g/t for transitional material. Centamin moved forward with Stage 3 of the project to increase mill throughput by 25 per cent to 5 Mt/yr from 4 Mt/yr, which is expected to be completed in 2011. The company began a scoping study to choose the best process flow route that would achieve a new mill throughput target of 10 Mt/yr. The result of the study indicated that the conventional semi-autogenous grinding/ball mill grinding configuration would meet the desired target. Centamin expected to complete the goal of the Stage 4 expansion plan of 10 Mt/yr of mill throughput in 2012 [Centamin Egypt Ltd., 2011,].

Gippsland moved forward with exploration works at eight gold prospects and one copper-nickel prospect at the Wadi Allaqi region, which is located 250 kilometers (km) southeast of Aswan in the Eastern Desert, Figure 14.2. Gippsland's drilling

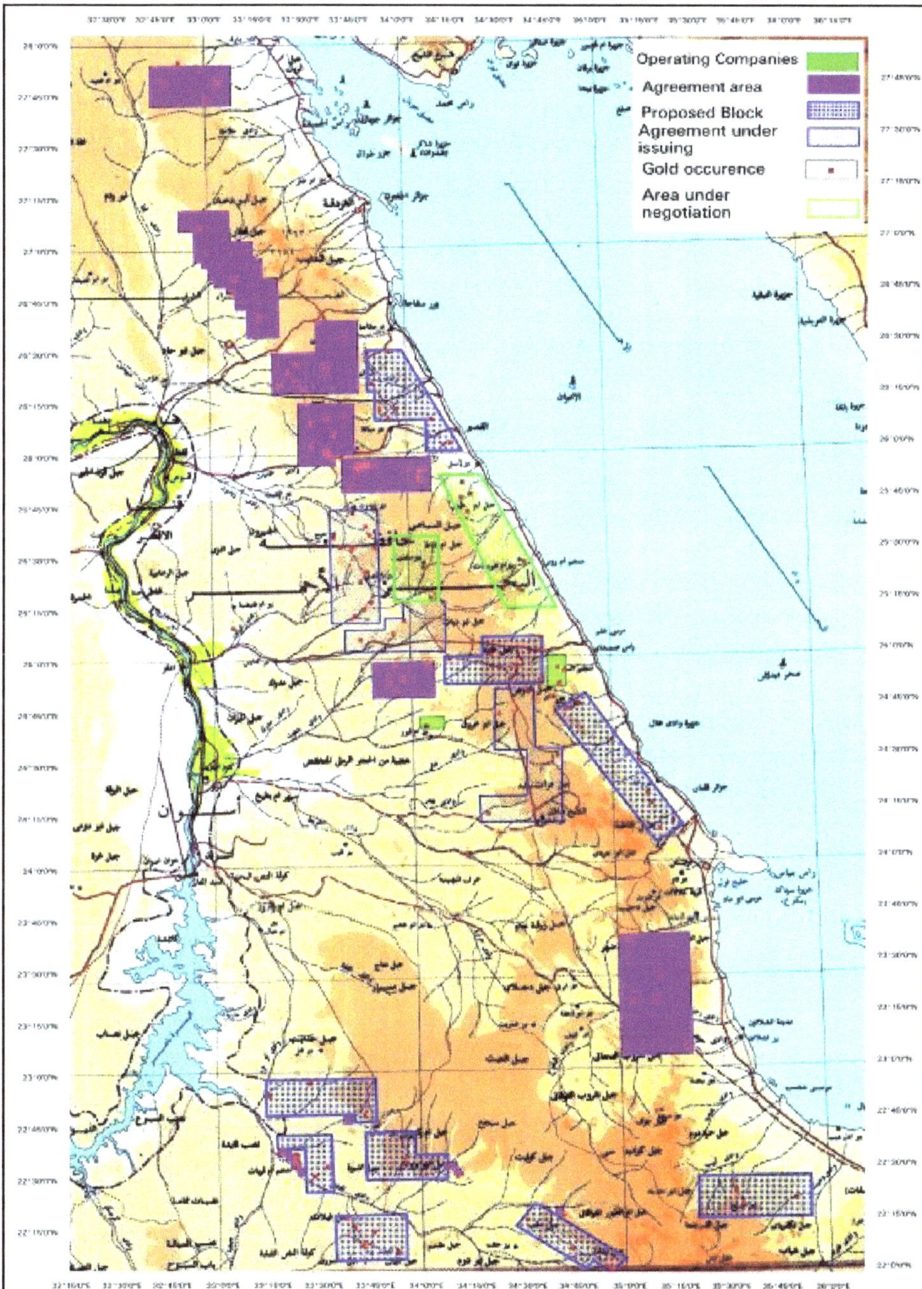

Figure 14.2: Location of Gold Deposits in Egypt [EMRA, 2006].

programme identified several mineralization areas, including at Seiga, which had an estimated inferred gold resource of 2,635 kg [Gippsland Ltd., 2011].

Hamash Gold Mine, which is located 100 km west of Marsa Alam in southeastern Egypt, produced 465 kg of gold in 2010 compared with 60 kg of gold in 2009. The mine used a heap-leach production method and was operated by Hamash Egypt for Gold Mines, which was a 50-50 joint venture between Cypriot Matz Holdings of Cyprus and EMRA [Ministry of Petroleum, 2011].

In February, EMRA awarded the joint venture of Nuinsco Resources Ltd. (NWI) of Canada and Quartz Core Ltd. (QCC) a gold exploration license at the Bukari and the Umm Samra concessions. NWI and QCC established a joint venture, Z-Gold Resources, to conduct exploration activity at the Bukari and the Umm Sammra concessions. The company finalized the Bukari concession agreement with EMRA and moved to finalize the Umm Samra concession agreement. The Bukari concession is located about 50 km northwest of the Sukari gold mine and has 18 known gold occurrences. Chemical assays of 66 samples from the Bukari concession contained an average of 3.35 g/t gold [Nuinsco Resources Ltd., 2011; Quartz Core Ltd. for Mineral Resources, 2011].

Alexander Nubia Inc. (ANI) of Canada held two exploration concessions in Egypt the Abu Marawat and the Fatiri concessions which are located in the Eastern Desert and together cover a total land area of 2,772 square kilometers. The Abu Marawat gold-copper-zinc-silver project was a concession that ANI acquired in an international competitive auction from Centamin in 2008. ANI drilled 79 diamond drill holes and planned to release estimates of the Abu Marawat's mineral resources in the first quarter of 2012 [Alexander Nubia Inc., 2011].

Recently, gold was discovered at Uwaynat area, G. Kamel and G. Nazar. The gold assay in this area is up to 14 g/t [Naim, 1998; Khalid, 1996; Said, 1998]. This area is recently bided for exploration and exploitation [EMRA, 2006]. There are Bids that were decided upon in 2007 for gold exploration and exploitation [EMRA, 2006]. These bids are for the locations of: Um Balad, El-Fawakhier, Fatiri, Abu Marwat, Wadi Kariem, Hodine, Dungash, Uwaynat, and Barramiya.

As a matter of fact, a second bid for gold exploration and exploitation has been announced to the public for additional areas in the Eastern Desert.

The Future of gold mining and processing in Egypt is a bright one. Extensive exploration, mineralogical, petro-graphic, and processing research work is necessary for profitable exploitation of the gold resources in Egypt.

Iron Ores

Iron ores in Egypt occur in two forms; banded iron formation (BIF) and ironstone.

Eastern Desert BIF

The most famous occurrences of BIF in Eastern Desert are Wadi Kareim, Um Nar, Abu Marawat, El Dabbah, Um Ghamis, Gabal El Hadid, Um Shadad, and Abu Diwan,[Hunting Geophysical Co., 1974].

Ironstone Deposits

Iron stone is the iron ore which is formed mainly within Phanerozoic sediments and is well represented in Egypt in Bahariya and Aswan iron ores.

Bahariya Iron Ore

Several iron ore deposits are located in Bahariya area, *e.g.*, El Harra, El Heiz, Ghorabi, El Gedida, and Nasser. The iron content in the ironstone deposits ranges from 30 per cent to 58 per cent Fe, and the manganese content ranges from 0.7 per cent to 7.66 per cent Mn [El Bassyony, 2000]. Only the exploitable iron ore in Bahariya Oases at El Gedida area has little or no overburden. Originally, when mining started in this area in 1972, the minable reserves were estimated accurately by 135 Mt. Today, the left minable reserves are estimated by only 63 Mt, which are just enough for about 15-20 years at the present mining rate of 3 to 3.5 Mt/y. The other areas: Ghorabi, Nasser, El Heiz, and El Harra are of low grade ores and of high manganese content. In addition, these areas have relatively thick overburden.

Aswan Iron Ore

Iron ore in this area was known since Pharaonic time. In recent years it was the main supply of iron ores for the Egyptian iron and steel industry till 1972 when it was replaced by Bahariya iron ore. According to Hussein [1990], the ore is a bedded oolitic type of Senonian age in the form of two bands interbedded with ferruginous sandstone and clay capping Precambrian rocks. The thickness of the bands varies from 0.2 to 3.5 m. The main iron minerals are hematite with minor goethite where quartz, gypsum, halite and clay are gangue minerals. The reserve was estimated between 121-135 million tones with average content of 46.8 per cent Fe [Attia, 1955]. The ore had been formed under sedimentary lacustrine conditions during Senonian sedimentation. Aswan iron ores were used to feed the steel plant at Helwan from its establishment in 1956 until 1972 when the Bahariya iron ore, from El Gedida area, started to replace Aswan ore in the iron and steel plant at Helwan.

The potentiality of discovering more iron ore is high especially in the area between Bahariya Oases in the North and Uwaynat area in the South on the bases of geological and structural observations. Reserves and production of iron ores in Egypt are shown in Table 14.1 [Abouzeid, 2008]. In this table, it is clear that Egypt is running short of the available indigenous exploitable iron ores, which is mainly in El Gedida area, Bahariya Oases, Western Desert. The Aswan iron ore is high in phosphorus content, in addition to its peculiar formation. The Eastern Desert BIF iron ores are of small quantities (about 50 million tons in total in all localities) spread in a vast area of about 200 km^2. Uwaynat BIF iron ore has been recently discovered and not thorough investigations (exploration, reserve estimation, characterization and/or beneficiation) have been carried out. In addition, the area is more than 600 km far from the inhabited Nile valley area, with little or no infra structure. A preliminary project for mining, processing, and pelletizing of Uwaynat iron ore deposit is being proposed after the discovery of Uwaynat BIF deposits [Abdel Meguid, 1992; Said, 1998; Khalid, 1996]. Pilot scale mineral processing tests showed that Uwaynat ore may be upgraded to 66 per cent Fe at a reasonable recovery [Khalid, 2002].

Bahariya Oases GedidaGhorabiNasserAl-HarrahAl-Haiz
90603035100 3,000—— 5347444530 Mn,S,P,Cl,SiP,Si
Hematite+Geothite+Limonite+hydrogeotjite

Table 14.1: Iron Ores information [Abouzeid, 2008]

Area	*Location*	*Reserves, M tons*	*Prod-uction 1000 t/y*	*Average Assay, Fe %*	*Associated Constituents*	*Remarks*
Aswan	N-E Aswan	25	20	44	Mn, S, P,	Hematite
	S-E Aswan	300		46	SiO_2, Ti	+hydrated
	West Aswan	na		44		oxides
Eastern Desert	W.Kareem	18	–	45	Si,Ti,P,Ca	Magnetite+
(BIF)*	Um Nar	14	–	46		Hematite
	Abu Marwat	7	–	44		
	W.Dabbah	6	–	38		
	Um Khamees	6	–	44		
	G. Elhadid	4	–	46		
	Abou Rakab	5.7	–	37		
	El-Hendousi	0.2	–	31		
	Um Shaddad	0.3	–	48		
	Sitro	0.55	–	43		
Bahariya Oases	Gedida	90	3,000	53	Mn,S,P,Cl,Si	Hematite+
	Ghorabi	60	–	47	P,Si	Geothite+
	Nasser	30	–	44		Limonite+
	Al-Harrah	35	–	45		hydrogeotjite
	Al-Haiz	100	–	30		
Uwaynat (BIF)*	G.Nazar	15-300 m	–	Dec-42	Si,Au,Ag,	Magnetite+
	G.Kamel	Thickness	–		Cu,Pb,Zn	Hematite+
	G.Arkenu	Several kms Extention	–			Silicates

Coal

The El-Maghara coal mine is the only producing coal mine in Egypt. It is located in the middle of the Sinai desert about 250 km north-east of Cairo, where a coal-fired power plant is intended to be built, Figure 14.3.

Maghara region is located nearly in the center of North Sinai between longitude 33°10′ and 33°40′ E, and latitude 30°35′ and 30°50′ N, covering an area of about

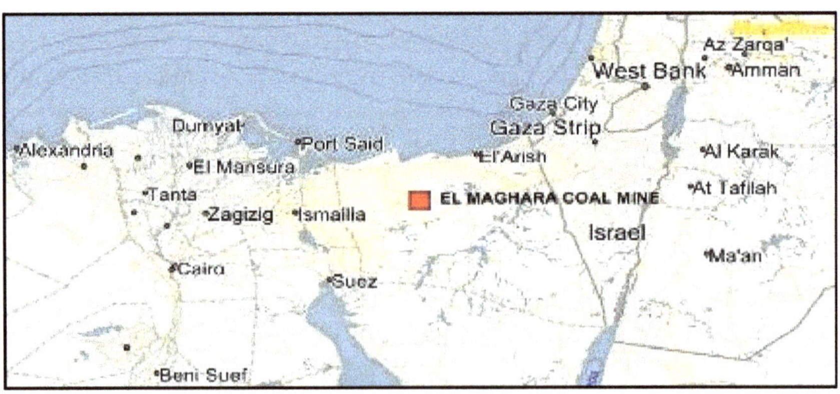

Figure 14.3: Location of el-Maghara Coal Mine, [www.infomine.com/index/elmaghara coal.html].

1,300 km². The region forms about 54 km long and 30 km wide asymmetric anti-cline with a curved axis trending about 55° NE and thus it is par-allel to the Syrian Arcs. The thickness of the Main Coal Seam MCS in the Maghara area ranges from 130 to 190 cm and the geologic coal reserve is estimated to be 52 Mt based on the drilling data of more than 50 exploration wells [Mostafa, 2001].

Proximate and ultimate analyses, calorific values and forms of sulfur are summarized in Table 14.2, [Melegy, 2009]. Total moisture contents are 2.65 wt. per cent for fresh coal (FC) and 4.45 wt. per cent for weathered coal (WC). Volatile matter contents are 50.6 wt. per cent in FC and 39.36 wt. per cent in WC. Fixed carbon contents are 40 wt. per cent in FC and 29 wt. per cent in WC. Proximate results indicate that the coal samples have a low average ash content of 4.12 wt. per cent in FC and a high average ash content 22.48 wt. per cent in WC.

Table 14.2: Proximate, Ultimate and Petrographic Analyses of Fresh and Weathered Coal in Maghara Mine, [Melegy, 2009]

Parameters	*Fresh coal sample (FC)*	*Weathered coal sample (WC)*
Proximate analysis		
Ash	4.12	22.48
Moisture	2.65	4.45
Volatile	50.6	39.36
Fixed carbon	40.00	29.00
Ultimate analysis		
Carbon	71.44	48.5
Hydrogen	6.62	5.18
Oxygen (by diff)	9.1	11.32
Nitrogen	2.27	1.56
Sulphur	3.8	6.15
Forms of sulphur		
S (total)	5.77	8.82
S (sulphate)	1.14	2.30
S (pyrite)	3.80	6.4
S (organic)	0.83	0.12
Calorific value(cal/g)	7422	4763
Petrographic analysis		
Vitrinite	91.6	41.4
Liptinite	7.5	11.1
Intenite	0.0	1.00
Minerals	0.9	46.5

Calorific values were 4763cal/g and 7422cal/g for WC and FC samples, respectively, placing most of the fresh coal samples in the high volatile bituminous coal. Maghara coals, in general, have high total sulfur values, ranging from 5.77 wt. per cent in FC to 8.82 wt. per cent in WC. Total sulfur is dominated by pyritic form.

Tantalum

Tin-Tantalum-Niobium deposits are genetically related to late phase of granitic intrusions which generally form small and simple intrusions of alkali to peralkali intra-plate anorogenic granites. Most of the Sn, W, Mo, Nb-Ta,REE,Be and F deposits are associated with this type. The geological, mineralogical, and geochemical characteristics of this mineralization let many authors to propose the metasomatic origin [Sabet, 1976]. The famous occurrences of this mineralization, which reach the economic level, are Abu Dabbab, Neuweibi, Muelha, Um Naggat, and Abu Rusheid.

Gippsland has agreed upon a joint venture to evaluate the Abu Dabbab tantalite deposit that is estimated to contain measured resources of 12 Mt grading 274 g/t tantalum pentoxide. Indicated and inferred resources total 28.1 Mt grading at an average of 250 g/t tantalum pentoxide. A new company, Tantalum Egypt is to be formed that represents the interests of both Gippsland and the EGMSA. The Abu Dababb deposit is located 770km south of Cairo and 75km from the Port of Quseir.

Gippsland intended to produce a high-purity tantalum synthetic concentrate, known as SynCon [a tantalum concentrate that contains 55 per cent tatantum pentoxide (Ta_2O_5)]. In addition to tantalum pentoxide, the Abu Dabbab Mine was expected to produce 1,530 t/yr of tin metal in concentrate, which would be sold on the open market. Commissioning of production at the Abu Dabbab Mine was scheduled for 2012. The Nuweibi site contained 98 Mt of tantalum-tin-feldspar mineral resources, which was double that of the Abu Dabbab deposit, but the average Ta_2O_5 grade was 40 per cent lower than that of Abu Dabbab. Given the size of identified resources of 142.5 Mt from the Abu Dabbab and the Nuweibi deposits, Gippsland expected to rank among the world's leading producers of tantalum for many years to come (Gippsland Ltd., 2011, p. 5).

Manganese Ores

Manganese ores occur in Egypt in two major localities beside other several small occurrences [Attia, 1956]. The economic deposits of manganese are Um Bogma in Sinai and Elba in South Eastern Desert.

Um Bogma area lies in Central Western Sinai Man-ganese occurs as lenses and lensoidal bodies with differ-ent dimensions within Carboniferous sediments of Um Bogma formation.

Elba Manganese ore occurs within sedimentary rocks of Miocene age in about 24 locations in Shalateen plain. Other Mn occurrences are recorded in Wadi Malik near Ras Banas and Abu Shaar El Qibli in Southern part of Esh El Mal-laha range [Said, 1990, Hussein, 1990]. The reserves and production are given in Table 14.3.

Manganese is mined in the different localities by underground methods, mainly room and pillar. In the exposed outcrops it is mined by surface mining techniques.The iron oxides are highly disseminated in the manganese oxide matrix, which makes the possibility of upgrading the ore is limited.

Table 14.3: Manganese Ores information [Hussein, 1990]

Area	*Location*	*Reserves, M Tons*	*Produc., 1000 t/y*	*Average Content, MnO_2*	*Associated Constituents*
Eastern Desert	Esh El-Mallahah	NA	NA	35.8-45.14	Fe_2O_3,
	W. Ma'alik	NA	120	42.17	SiO_2,
	G.Elba and Abou Ramad	NA	NA	45.0	clays
Sinai	Abou Zunima	5	120	38	Fe_2O_3, SiO_2, clays

The only processing steps carried out on the manganese ores are crushing and screening. The prepared ore is mixed with some imported high grade manganese ore and fed to the smelter at Abu Zunaima, Sinai to produce ferromanganese alloys [ESAC, 2007]. The fines, under size fractions, are piled in dump areas. It is expected that agglomeration and magnetic roasting, followed by low intensity magnetic separation may improve the grade of the manganese ore in the rejected fines, which in turn may increase the manganese recovery.

Sulphur

Sulphur deposit is located in the area of Degla, figure (4), east of Ariesh City north of Sinai,[El Sayed,1989; John,1988]. Sulphur formation extends for about 400 km^2 on the surface and 250km^2 underground. In 1985, Free Port Company has studied sulphur formation in an area of 13 km^2 and a reserve of 20 million tons of sulphur with a high grade was estimated. The feasibility study of the project gave negative results, at that time, due to the lower price of sulphur and high price of the required energy for the project. Nowadays, Egypt import about ½ ton of sulphur which costs about 30 million $. This will increase to one ton in the next four years. However, it is recommended to reevaluate the sulphur project due to the presence of natural gas line in the Ariesh city as a source for the energy needed for the project and the improvement in the sulphur price, Table 14.4.

Table 14.4: Accessibility of Degla Region for Sulphur Project

Parameters	
Area	13 km^2
Locality	Digla region, East El Arish City - N. Sinai
Lat.	31º 06ó 17.4" 31º 10ó 39" N.
Long.	34º 00ó 33.6" - 34º 02ó 51" E
Nearest Cities	El Arish (23 km East), El Sheikh Zuwayid (8.5 km NE)
Accessibility	The area can be easily reached through the main Asphaltic Road of El Arish – Rafah.
Energy sources	The gas pipe lines nearest the area.
Ports	El Arish Port

Contd...

Table 14.4–Contd...

Parameters	
Prospectivity:	
Previous work	Freeport Egyptian Sulphur Company
Host Rock	Evaporates unit (Sulphur bearing unit
Mineralization	Gypsum and anhydrite, altered limestone, and altered limestone with Sulphur Ore.
Ore Grade	Average 17 per cent
Mining Method	Frasch Process method
Thickness	18 m
Geologic Reserve	20 million Tones

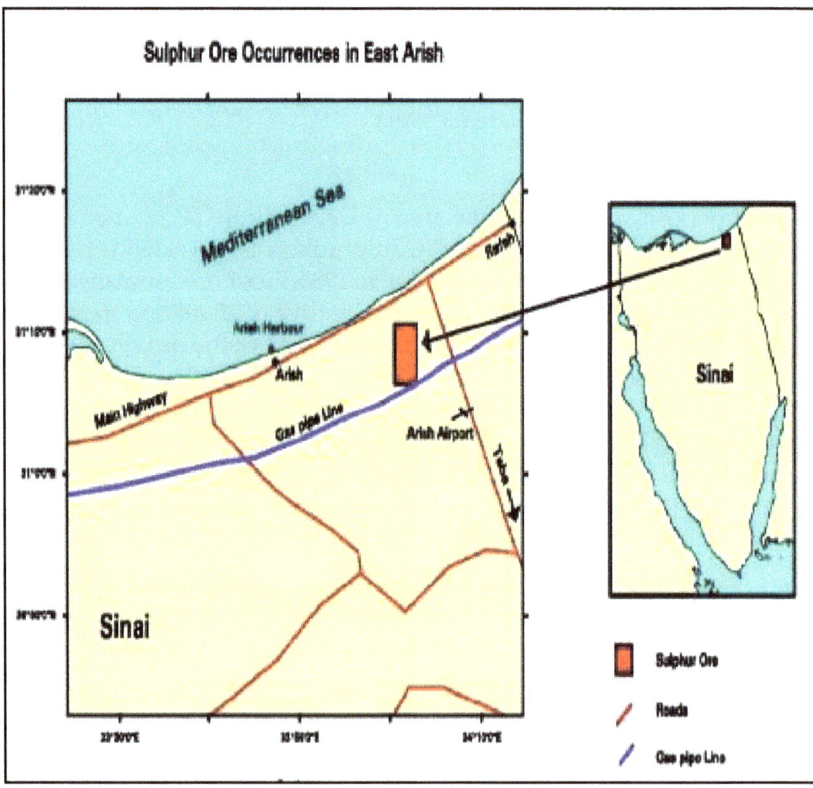

Figure 14.4: The Degla Region is Potentially Commercial Deposit [El Sayed, 1989].

Ilmenite

The main source of titanium in Egypt is ilmenite. Ilmenite is present in a rock form in different localities in the Eastern Desert, and in the black sands on the Eastern part of the Mediterranean Coast, Figure 14.5 [Abouzeid, 2008; Naim, 1993]. Titanium ores information is presented in Table 14.5.

LEGEND

● Hard rock deposits

1-Abu Ghalaga 2- Abu Dahr

3- Umm Effin 4- Umm Tundub

5- Wadi El Miyah 6- Koimnab (El Hamra Dome)

○ Placer, deposits (Black sands)

1- Rosetta 2- Damietta

3- El Arish

Figure 14.5: Ilmenite Localities in Egypt.

Table 14.5: Titanium and Titaniferous Iron Ores [Abouzeid, 2008; Naim, 1993]

Area	*Reserves, M tons*	*Produc., 1000 t/y*	*Average Assay, TiO_2 per cent*	*Associated Constituents*
Red SeaCoast	40	120	30 - 38	Fe_2O_3, SiO_2, Clays, Silicates
Mediterranean Coast (Black sands)	400	–	2 - 3	SiO_2, Mag., Rutile, Zircon, monazite

In Egypt, abundant reserves of titanium bearing minerals are available as ilmenite, which is estimated to be about 40 million tons as igneous rocks at several places in the southern part of the Eastern desert, Figure 14.5. The chief ore bodies occur in Hamata region of El-Nasr Mining Company at Abu-Ghalaga. In addition, titanium bearing minerals occur as an essential constituent of beach placer, called Black Sands. So far, no industry has been established in Egypt based on these resources. The technical know-how for the production of titanium metal is restricted to a group of developed countries. The detailed studies given by Hussein [1990] revealed the presence of three types of ilmenite ore:

1. Red ore or oxidized zone on the surface,
2. Black ore or the main body, and
3. Disseminated ore

The mineralogical studies showed that the ore contains the following minerals:

Ilmenite 67.4 per cent - 68.8 per cent, Hematite 13 per cent - 18 per cent, Secondary hematite 15 per cent, Pyrite 0.13 per cent - 2.1 per cent, other minerals 4 per cent - 11 per cent. The overall chemical analysis of the ore is presented in Table 14.6.

Table 14.6: The Overall Chemical Analysis of the Ore

Oxide	*Oxide oxidized zone*	*Fresh ore*
TiO_2	37.09 per cent - 41.04 per cent	33.9 per cent - 37.65 per cent
Fe_2O_3	17.47 per cent - 23.0 per cent	6.34 per cent - 23.85 per cent
FeO	27.93 per cent - 35.63 per cent	25.94 per cent - 31.33 per cent
V_2O_5	0.3 per cent 1 - 0.38 per cent	0.29 per cent - 0.39 per cent

The Abu Ghalaqa ore is being mined by surface mining. The benches are drilled, charged, and blasted. To facilitate loading, transporting, and crushing, secondary blasting is applied on the oversize boulders.

Black Sands

Black sands in Egypt are beach placers deposited from the Nile stream during flood seasons reaching the Medi-terranean Sea at river mouth. It spreads on the beach East of Rashid branch of the Nile and extends east to Rafah passing through El Arish coastal plains [Hassan, 2003]. Figure 14.6 shows the geographic distribution of the black sands in Egypt. They spread along the Mediterranean Sea shore from Alexandria West to Rafah East. The black sands contain some economic minerals such as ilmenite, hematite, rutile, magnetite, zircon, garnet, and monazite. According to Naim *et al.*, [1993], the reserves of economic minerals at Rashid area are presented in Table 14.7.

Table 14.7: Reserves of Economic Minerals at Rashid Area, Naim *et al.* [1993]

Mineral	*Reserve, (in 1000 tons)*
Ilmenite	2087
Magnetite	1437
Hematite	214
Zircon	81
Rutile	29
Garnet	72
Monazite	31
Sulphides	86
Heavy silicates	1315

Role of CMRDI in Maximizing the Benefits of Minerals Resources

High demands are made on the raw materials with respect to homogeneity and stability of composition; no impurities are permitted. Specifications of raw materials for certain industrial applications are, sometimes, very stringent in terms of particle size, chemical analysis, color, brightness and viscosity. However, suitable beneficiation processes for these materials are needed to satisfy the market specification. Central Metallurgical Research and Development Institute (CMRDI) is a nonprofit governmental institution established officially in 1983 in the industrial area of El-Tabbin, Helwan, in the south of Cairo, Egypt. Before 1983 and since the beginning of the sixties, CMRDI was a part of the National Research Center (NRC) at Cairo. CMRDI is the main institute in Egypt to be referred to when issues of mineral resources and mineral industries are to be discussed. In fact, CMRDI played, since its foundation, a major role in maximizing the benefits of the minerals resources not only in Egypt, but also in some Arab and African countries as well.

The mission of the institute is to contribute to the national economy and the welfare of the society by improving the Egyptian industry through technological development and technology transfer. CMRDI, supported by the well developed infrastructure, was able to play an effective role in transferring many technologies to the Egyptian industries as well as to Africa and Middle East countries. In the area of evaluation, beneficiation and processing of minerals and ores, CMRDI has a multiplicity of experience of over 30 years. Through three decades CMRDI was able to build a unique international school in such areas of research where the senior staff got their PhDs' from different developed countries such as United States of America (USA), Japan, England, France, Former Soviet Union (FSU), Poland, Romania, and

Germany in addition to the Egyptian universities. So, the experts of CMRDI are able to help the industry in maximizing the benefits of the minerals wealth. Specifically, CMRDI evaluated and upgraded most of the Egyptian ores and minerals such as iron ore, phosphates, kaolin, quartz, silica (white) sand, black sand, ilmenite, feldspars, coal, petroleum coke, limestone, dolomite, bentonite, clays, barite, talc, chromite, manganese, zinc-lead ores and others. These studies are performed on different levels starting from the bench scale level to test the different available technologies to upgrade and process the ores and suggesting a suitable flowsheet, to a pilot scale level, with taking into consideration the techno-economic and environmental aspects required to determine their viability. CMRDI has 2 pilot-plants for treating minerals. These pilot-plants are multi-purposes and integrated together in such a way that they can deal with any mineral commodity starting from beneficiation till producing finished value- added products.

Successful Examples for Maximum Utilization of Minerals in Egypt and some Arab and African Countries

CMRDI evaluated and upgraded most of the Egyptian ores and minerals such as iron ore, phosphates, kaolin, quartz, silica (white) sand, black sand, ilmenite, feldspars, coal, petroleum coke, limestone, dolomite, bentonite, clays, barite, talc, chromite, manganese, zinc-lead ores and others.

In the Arab region, CMRDI conducted several successful projects for evaluating the iron ores in Syria and Libya in addition to recover the valuable metals from some metallurgical wastes in Saudi Arabia. On the African level, CMRDI evaluated the mineral industry in Nigeria.

Meanwhile, CMRDI has longstanding successful collaboration with many national and international institutions in the fields of mineral resources, mineral beneficiation, and mineral processing industries.

The projects are conducted on different levels as follows:

- *Bench scale level*: these studies aim to evaluate the different available technologies to beneficiate and process the ores or minerals and to suggest flow sheets that are technically viable.
- *Pilot scale level*: this aims to perform the experiments on a continuous base to generate all the techno-economic data required to scale-up, machine selection as well as preparing the basic engineering and the pre-feasibility study.
- *Plant testing*: the pilot plant studies also aims to produce sufficient quantity of mineral concentrates to be tested by the end-user in the plant.
- *Techno-economic feasibility*: based on the above-mentioned studies, the techno-economic feasibility is prepared.
- *Technology transfer*: CMRDI can help in transfer of technology to industry. This can includes, among others, helping in machine and offer selection, training engineers and technicians, and technical support.

The following are some selected projects in the topic of utilization of mineral resources:

In Egypt

- Evaluation and beneficiation of Egyptian iron ores.
- Studies for evaluation of local bentonite ores for industrial and agricultural applications.
- Evaluation of Egyptian phosphate ores for production of phosphoric acid and phosphatic fertilizers.
- Economics and future of Egyptian phosphates.
- Utilization of Egyptian kaolin ores in sanitary ware production to replace the imported kaolin.
- Study the economics of the Egyptian phosphate ores and their future in the light of local and international markets.
- Survey and study of mineral resources used in glass and ceramic industries.
- Abu-Tartur phosphate project: An outlook in view of the current situation of the project and the international phosphate Market.
- The National Project for Upgrading the Egyptian Ores Required by the Local Industry.
- Desulphurization of El-Maghara Coal by Physical and Mechanical Techniques.
- "Beneficiation of Some Local Glass Sands.
- Evaluation of Egyptian calcium carbonate as a filler for application in plastics, pigments and paper industries.
- Pilot plant production of refractory grade chromite concentrate.
- Production of silicon and its carbide from local ores and fumed silica of ferro-silicon plants.
- Survey and study of ores used in glass and ceramic industries: white sands and feldspars.
- Beneficiation of low grade and fine Sinai manganese ores for ferromanganese and battery grade manganese production.
- Separation of zircon from beach sands.
- Beneficiation of Egyptian Talc-Magnesite ores.
- Production of magnesium sulfate salts from Egyptian magnesite ores.
- Production of barium salts from Egyptian barite ores.

In African Countries

Libya

- Mineral beneficiation and processing in collaboration with Industrial Research Institute, Tripoli.

- ✰ Assessment of Wadi Shatei iron ores for iron making industry by blast furnace or direct reduction techniques.

Nigeria

- ✰ Development of a plan for maximum utilization of local natural mineral resources, Raw Materials Research and Development Council, RMRDC, Lagos.

Tanzania

- ✰ JICA-Egypt assignment to TIRDO for mineral beneficiation development programme.

In Asian Countries

Syria

- ✰ Evaluation of iron ore deposits of Rago and Zalabany localities – Department of Geology.

Outlook

The Egyptian mineral industry has enormous growth potential for the next decade owing to the country's abundant mineral resources—especially gold, natural gas, phosphate rock, and tantalum—and the increased demand for steel reinforcing bar and cement by the construction sector. Domestic demand for cement and steel is expected to increase in the short term as 524,000 low-cost residential units are being built throughout the country and expected to be completed by 2012. Egypt expects to be a major producer of gold and tantalum in the next 5 years. The passage by the newly elected Parliament of the proposed mining law, which was drafted with the help of the World Bank, could reduce bureaucracy, enhance transparency, and facilitate financing to mineral industry projects.

Possible Areas for Investment in Mineral Industry in Egypt

The following areas are open for serious investment in the mineral industry, in Egypt:

1. Mining and Mineral Processing of iron ores at: Uwaynat (Western Desert), Eastern Desert, Bahariya Oases, and Aswan for iron pellets production.
2. Exploitation of ilmenite ores in the feasible areas.
3. Exploration, Mining, Processing, and Extraction of: gold, tin, tantalum, and niobium.
4. Exploitation of manganese ores.
5. Exploitation of sulphur.

References

1. Abdel Fattah, M. F., 2005, "Petrological and Geochemical Studies on the Basement Rocks of Northeast Gabal Uwaynat Area, Western Desert, Egypt," M. F. PhD Thesis Suez Canal University, p. 245.

2. Abdel Meguid, A. A., 1992, "Late Proterozoic Pan African Tec-tonic Evolution of the Egyptian Part of the Arabian-Nubian Shield," *Middle East Research Centre*, Ain Shams University, v.6, pp.13-28.Taib, M., The Mineral Industry of Egypt, U.S. Geological Survey Minerals Yearbook-2010.
3. Abouzeid, A.Z. M. and El Wgeeh, M. A., 2008, "Mineral Indus-try in Egypt-State of the Art," 11*th International Mineral Processing Symposium*, Belek, Antalya, Turkey., pp. 1-27.
4. Abu Salem, A. O., 2003, "Geology and Mineralogy of the Basement Rocks of West Gabal Kamel area," M.Sc.Thesis, Al Azhr University, p. 90.
5. Alexander Nubia Inc., 2011, Abu Marawat project: Alexander Nubia Inc. (Accessed January 25, 2012, at http://www.alexandernubia.com).
6. Arab Fertilizers Association, 2010, Members—New Valley phosphate project: Arab Fertilizers Association. (Accessed November 15, 2010).
7. Arab Fertilizers Association, 2011, Fertilizer statistical yearbook 2010: Cairo, Egypt, Arab Fertilizers Association, p. 64.
8. Arab Investment and Export Credit Guarrantee Corp., 2011, Arab Republic of Egypt: Arab Investment and Export Credit Guarrantee Corp., 2 p. (Accessed October 23, 2011, at http:/www.iaigc.net/User Files/file/ar/Country Risk/Egypt.pdf)
9. Attia, M. I., 1955, "Topography, Geology, and Iron Ore of the District East of Aswan," *The Egyptian Geologic al Survey*, p. 262.
10. Attia, M. I., 1956, "Manganese deposits of Egypt," *The 20th International Geological Congress*, Mexico, pp. 143-171.
11. Centamin Egypt Ltd., 2011, Annual report—31 December 2010: Centamin Egypt Ltd., 104 p. (Accessed January 26, 2102, at http://www.centamin.com/wp-content/uploads/2011/03/Centamin-Annual-Report-31-Dec-2010.pdf.)
12. Central Bank of Egypt, 2011, Annual report 2010: Cairo, Egypt, Central Bank of Egypt.
13. Egypt Economy 2012, 2012 World FACT Book CIA and other sources.
14. El Bassyony, A. A., 2000, "Geological Setting and Origin of El Harra Iron Ores, Bahariya Oases, Western Desert, Egypt," *Annal of Geological Survey of Egypt*, Vol. 23, pp. 213-222.
15. El Sayed A.A.Youssef, 1989, Geology and genesis of sulfur deposits at Ras Gemsa area, Red Sea coast, Egypt *Geology, September*, v. 17, p. 797-801.
16. EMRA, 2006, "Egyptian Mineral Resources Authority," *Results of* 2006 1*st International Bid Round for Gold Exploration and Exploitation in Egypt*, Egypt.
17. ESAC, 2007 "Egyptian Steel Alloys Co. Projected Plan for 2007-2008," pp. 1-46.
18. Gippsland Ltd., 2011, Annual report: Claremont, Australia, Gippsland Ltd., June 30, p. 80.

19. Hassan, M. A. "Black Sands Project," *A Briefing to the Egyptian Association for Mining and Petroleum, Nuclear Material Authority*, Cairo, June 12, 2003, p. 21.

20. Hunting Geophysical Co., 1974, Geology of Al Uwaynat, East. Libya, IRC Tripoli. Internal Report, p. 190.

21. Hussein, A. 1990, "Mineral Deposits," In: R. Said, Ed., *The Geology of Egypt*, Taylor and Francis Publishers, Lon-don, pp. 511-566.

22. John L.S., Robert C.T. and Steven S., 1988, Sedimentological evidence for early Miocene fault reactivation in the Gulf of Suez, Geology, February, v. 16, pp. 113-116.

23. Khalid, A. M. and Diaf, A. A., 1996, "Geological and Geochemi-cal Exploration for Gold and REE at Jabal Nazar and Ja-bal Arkenu, Egypt-Libya," *Proceedings of Geological Survey, Egypt*, pp. 425-446.

24. Khalid, A. M., Greiling, O. R., Said, M. M., Megahed, A., Shaaban, G., Micheal, M. *et al.*, 2002, "South Western Desert BIF. Laboratory Studies and Gold Extraction Tests," *An-nal of Geological Survey of Egypt*, Vol. 25, pp. 315-332.

25. Khattab, M., Greiling, O. R., Khalid, A. M., Said, M., Kontany, A., Abu Salem, A. *et al.*, 2002, "Uwaynat Banded Iron Formation (SW Egypt) Distribution and Related Gold Mineralization," *Annals of the Geological Survey of Egypt*, Vol. 25, pp. 343-364.

26. Melegy,A. andSalman,S., 2009,Geochemical Studies on the Abandoned Maghara Coal Mine, Geoliness, 22, pp. 44-51.

27. Ministry of Petroleum, 2011, Companies investing in mining sector: Cairo, Egypt, Ministry of Petroleum. (Accessed January 15, 2011, (Accessed January 15, 2011, at http://www.petroleum.gov.eg/InvestCompanies.aspx.)

28. Mostafa, A.R. and Younous, M.A., 2001,Significance of Organic Matter in Recording Paleo Environmental Conditions of the Safa Formation Coal Sequence,Maghara Area, North Sinai,Eghypt, Int. Coal Geol,47, pp.9-21.

29. Naim, G., El Melegy, E. T. and El Azab, A., 1993, "Black Sand Assessment," *The Egyptian Geological Survey*, p. 67.

30. Naim, G., Khalid, A. M., Said, G. M., Shaaban, S., Hussein, A. and M. El Kady, 1998, "Banded Iron Formation Discovery at West Gabal Kamel and Its Gold Potentiality, Western Desert," *Annals of the Geological Survey of Egypt*, Vol. 21, pp. 303-330.

31. Nuinsco Resources Ltd., 2011, Egypt licenses: Nuinsco Resources Ltd. (Accessed January 25, 2012 at http://www.nuinsco.ca).

32. Quartz Core Ltd. for Mineral Resources, 2011, QCC in Egypt: Cairo, Egypt, Quartz Core Ltd. for Mineral Resources, p. 1.

33. Richter, A., 1986, "Geologie der Metamorphen und Magma-tischen Gesteine in Gebiet Zwischen Gebel Uwaynat und Gebel Kamel, SW Agypten, NW Sudan," *Berl. Geowiss. Agh*, Vol. 73, pp. 1-201.

34. Richter, A. and Schandelmeier, H., 1990, "Precambrian Base-ment Inliers of Western Desert, Geology, Petrology, and Structural Evolution," In: R. Said, Ed., *Geology of Egypt, Taylor and Francis*, pp. 185-250.

35. Sabet, A. H., Tsogoev, V. B., Baburin, L. M., Riad, A., Zakhari, A. and Armanious, L., 1976, "Geologic Structure and Laws of Localization of Tantlum Mineralization at Neweibi Deposit," *Annal of Geological Survey of Egypt*, Vol. 6, pp. 119-156.

36. Said, R., 1990, "The Geology of Egypt," Taylor and Francis Publishers, London,.

37. Sloss, L. L., 1979, "Global Sea Level changes: A View from the Craton," *Geological and Geophysical Investigations of Continental Margins, A. A. P. G. Memorial*, Vol. 29, pp. 461-467.

38. Said, M., Khalid, A. M., El Kady, M., Abu Salem, A. and Ibrahim, S., 1998, "On the Structural Evolution of Banded Iron Formation of Gabal Kamel and Its Role in the Gold Min-eralization," *Annal of Geological Survey of Egypt*, Vol. 21, pp. 345-352.

39. Van Houten, F. B., 1980, "Latest Jurassic-Early Cretaceous Re-gressive Facies," In: A. A. Craton and P. G. Bull, Eds., *Northeast Africa*, Vol. 64, pp. 857-867.

Chapter 15

Exploration and Potential Exploitation of Deep Sea Minerals in the EEZ and ECS of Mauritius

Kishore Boodhoo and Roshan T. Ramessur

Department of Chemistry,
University of Mauritius, Reduit, Mauritius
E-mail: kishore.boodhoo@uom.ac.mu

ABSTRACT

Mauritius, located in the South Western Indian Ocean, is a tiny island, yet is a huge ocean state with an exclusive economic zone (EEZ) of about 1.9 million km^2 and an extended continental shelf (ECS) is of about 396,000 km^2 respectively. Over this large extent of region lies deep sea minerals such as manganese nodules and marine polymetallic deposits found more precisely near ridges. Explorations are currently being conducted along segments of the Central Indian Ridge in the Indian Ocean by Korean and German teams in search for metalliferrous deposits. There is a need to develop a viable and sustainable marine mineral industry in Mauritius and to develop the sustainable management of deep sea mineral resources with improved human, technical and effective monitoring systems for exploration in the ECS.

Keywords: EEZ, ECS, Mauritius, Deep sea minerals.

Introduction

Approximately 71 per cent of the planet's surface (~3.6x10^8 km^2) is covered by oceans, known to be a vast storehouse of natural resources, hence capable of providing alternative sources of raw materials, for example metals that can be retrieved from metal deposits lying deep down into the sea. To successfully develop a mining industry a number of factors have to be taken into account namely technological advances, economical, legal and institutional frameworks. The oceans are made up of two distinct domains; the continental shelves and the abyssal plains. According to UNCLOS (United Nations Conference on the Law of the Sea), the continental shelf of a coastal State comprises the submerged prolongation of the land territory of the coastal State-the seabed and subsoil of the submarine areas that extend beyond its territorial sea to the outer edge of the continental margin, or to a distance of 200 nautical miles where the outer edge of the continental margin does not extend up to that distance (Mauritius Oceanography Institute, MOI, Mauritius).

An abyssal plain is an underwater plain on the deep ocean floor, usually found at depths between 3000 and 6000 m. Lying generally between the foot of a continental rise and a mid-ocean ridge, abyssal plains cover more than 50 per cent of the Earth's surface on the deep ocean floor (Wikipedia).

EEZ and ECS of Mauritius

The limits of the Exclusive Economic Zone (EEZ) of Mauritius (1850 km^2, 20°S, 58°E, South Western Indian Ocean) have been defined by geographical coordinates through subsequent regulations under the Maritime Zones Act 1977 as the Maritime Zones Regulations 1984. Mauritius has proclaimed its Territorial sea (12 nm) through the Territorial Sea Act of 16 April 1970, its Exclusive Economic Zone (200 nm, about 1.9 million km^2) around Mauritius and has also defined its continental shelf through the Maritime Zones Act 1977 (Figure 15.1). In addition, after making a joint submission

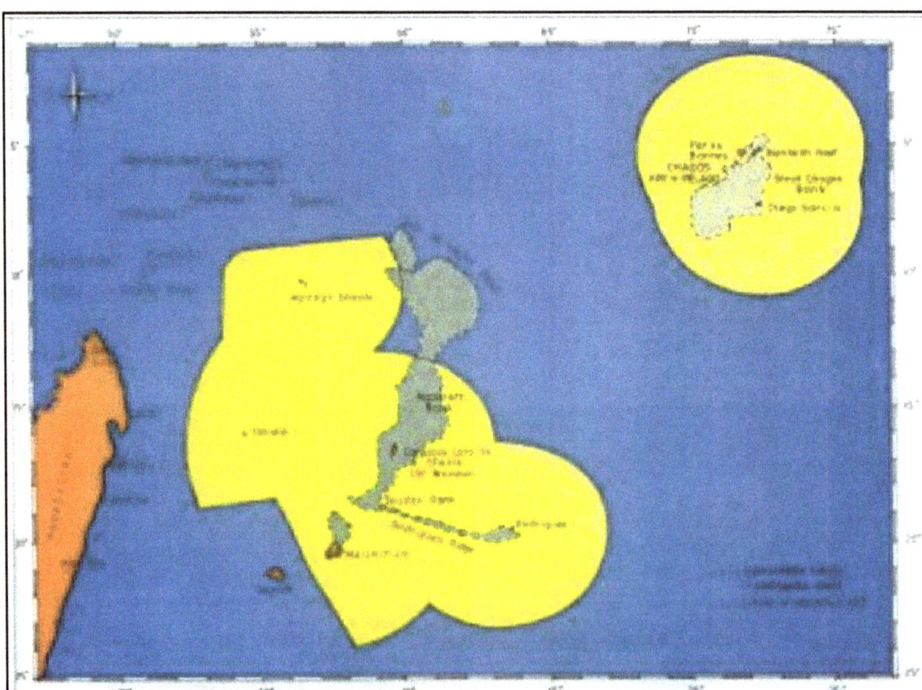

Figure 15.1: EEZ of Mauritius (MOI, Mauritius).

both by the Republic of Mauritius and the Republic of Seychelles to the United Nations Commission on the Limits of the Continental Shelf (CLCS) for an Extended Continental Shelf (ECS), CLCS finally made its recommendations in March 2011 such that Mauritius and Seychelles are to be conferred the jurisdiction upon the joint area of ECS of 396,000 km^2. This ECS (Figure 15.2) represents nearly 20 per cent of the size of the Exclusive Economic Zone (EEZ) of Mauritius (approximately the size of Germany).

Continental and Deep Sea Mineral Resources of the Oceans

The different deposits that can be recovered from the oceans are as follows (Yates, 1986):

Aggregates

These are deposits of sand, gravels and shells that were first recovered from the sea (Mero, 1978). They are used primarily in the construction industry, being a low priced commodity.

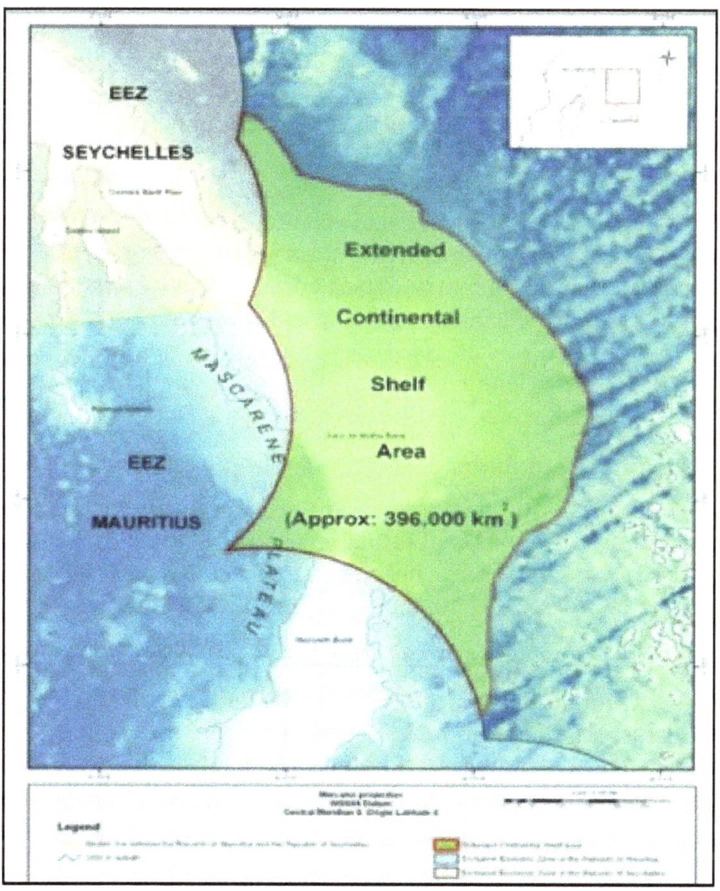

Figure 15.2: Joint ECS of Mauritius and Seychelles.

Placer Deposits

Placers occur as a result of chemical and physical weathering forming concentrations of minerals which are transported by running waters up to 15 km, with the larger, denser grains found near to the source. They contain a variety of valuable metals and gems. Japanese firms have mined magnetite in iron sands in Tokyo Bay and the Philippines (Cameron, 1983). Thailand had been planning to exploit tin deposits at a depth between 40 to 50 m. Marine placer mineral deposits occur and are widely exploited throughout the world, especially in Alaska, Florida, South Carolina, South Africa, South East Asia, Sri Lanka and UK.

Phosphorites

The principal mineral of marine phosphorite is carbonate fluorapatite or francolite. They exist as pellets, nodules, or sand at a depth of up to 1000 m where the nutrient-rich waters upwell to produce a wealth of biological activity. Significant deposits have been spotted at many offshore locations including South

West Africa, Florida in the US and Chatham Rise off New Zealand (Phosphorite nodules: Fertilisers from the Sea, 1984). Phosphate rock mined can be processed for fertilizer.

Metalliferrous Muds

Metalliferrous sediments, in which metals are concentrated, are common occurrences along all actively spreading ocean ridges and volcanic island arcs. The largest deposits were found in the Red sea whose submarine basin is known as Atlantis II Deep (Frazer, 1986). It is located at latitude 21° 23′ N and longitude 30° 04′ E. It has a surface area of 60 km^2 and occurs at a depth of 2200 m and acts as a trap for metalliferrous sediments owing to undersea volcanic activity and dense hot brines. The deposits can provide 920000 tons of zinc, 182000 tons of copper, 1635 tons of silver and 13 tons of gold (Mustafa, 1984). Segments of the Ridge closest to Rodrigues hotspots are 130-160 km in length typical of faster spreading rates than those further away are rich in metalliferrous deposits (Dyment, 2001; Tyler, 2003).

Marine Polymetallic Sulphides

The combined processes of volcanic activity, tectonic fracturing and hydrothermal circulation form massive deposits of sulphides at more than 2000 m depth. Seawater percolates down through fractures, becomes super-heated and leaches metals from the surrounding bedrock. Extremely volatile metal-rich solution is then ejected at temperatures of up to 350° C and mixes with cold alkaline seawater at 1° to 2° C. It rapidly precipitates out the metals it contains in the form of sulphides. This precipitation is so rapid that it produces an effect known as 'smoking' with dense clouds of particulate sulphides forming chimneys up to 20 m high at the mouth of the hydrothermal vents. The chimneys themselves eventually collapse, forming mounds of sulphides. Samples from various sites have contained up to 55 per cent zinc, 6.5 per cent copper and economic concentrations of silver (Marine Technology Society Journal, 1982).

Manganese Nodules

Of all these deposits, we are more interested with manganese nodules as well as metalliferrous sediments. It has been reported that there are cobalt rich manganese nodules lying in the exclusive economic zone of Mauritius (Fagoonee *et al.*, 1985). The EEZ of Mauritius comprises an area of about 1.9 million km^2. In addition, due to the joint submission made by the Republic of Mauritius and the Republic of Seychelles to the United Nations Commission on the Limits of the Continental Shelf (CLCS), an extended continental shelf of about 396,000 km^2 was conferred to both parties to exploit together of any resources available within this area.

They are encountered at a depth of about 5000 m. They have physically varied forms. They may look like potatoes, charcoal briquettes, cluster of grapes, cauliflowers, mushrooms and cannon balls. The different shapes reflect the different formation processes which are not fully understood. Manganese nodules also come into many sizes, in the range 0.5-10 cm in diameter, although most of them tend to be about 3-4 cm in diameter. Nodules also exhibit heterogeneous chemical compositions as shown in the Table 15.1 (United Nations, 1982):

Table 15.1: Different Composition of Manganese Nodules

Sample	*Mn (wt %)*	*Fe (wt %)*	*Co (wt %)*	*Ni (wt %)*	*Cu (wt %)*
Nodule A	9.50	17.00	0.16	0.13	0.07
Nodule B	25.60	4.10	0.16	1.57	1.28
Nodule C	18.63	19.64	0.48	0.31	0.13
Nodule D	42.30	2.47	0.17	0.26	0.15

The terms nodule abundance or nodule density or nodule concentration are applied as a measure of the amount of nodule on the seafloor, expressed in wet kg/m^2 (Holser, 1976). Assay or element concentration is a measure of the amount of an element present in a given sample, usually expressed as a percent of the total sample on the basis of dry weight. The most common methods for analysis of manganese nodules are atomic absorption, X-ray fluorescence and X-ray energy spectrometry. Grade is the content of an element of commercial interest. The sum of copper plus nickel content is commonly used in nodule resource literature to denote grade. Some authors have expressed grade as a 'nickel equivalent', taking into account, the different market values of the metals. For instance, a 'nickel equivalent' grade might be expressed as Ni + Cu/3 if the value of nickel is 3 times that of copper.

It is the Clarion-Clipperton Zone that proves to be the most potential site for manganese nodule mining. It is an area between the Clarion Fracture zone and the Clipperton Fracture Zone. The Clarion fracture zone is approximately 4000 km in length. It originates near the Line islands at 160° W longitude and terminates near the continental margin of southern Mexico at about 115° W longitude. The Clipperton Fracture is about 4500 km in length, originating at about 155 ° W longitude and terminating near the continental margin of Central America at about 105° W longitude. The fractures are roughly parallel.

Potential Mining Sites in the Indian Ocean

Fraser and Wilson had identified five potential sites for exploiting these manganese nodules in the Indian Ocean (Fraser and Wilson, 1979). The data collected revealed that the nodule grade is relatively constant over large distances, but more data are required to determine the tonnage of a module deposit (Frazer, 1979). Information on both grade and abundance are still insufficient for reliable estimates of the amounts of manganese, copper, and cobalt that might be recovered from the Indian Ocean. To be economically minable, the deposits should have at least an abundance of 5 kg/m^2.

An average grade of 2.47 per cent with a cut-off grade 1.8 per cent has been established as a criterion for paramarginal resources, also referred to potential reserves (Kildow, 1976). To be considered as a submarginal resource (minable within the next century) a nodule should have at least an average grade of about 1.6 per cent (United

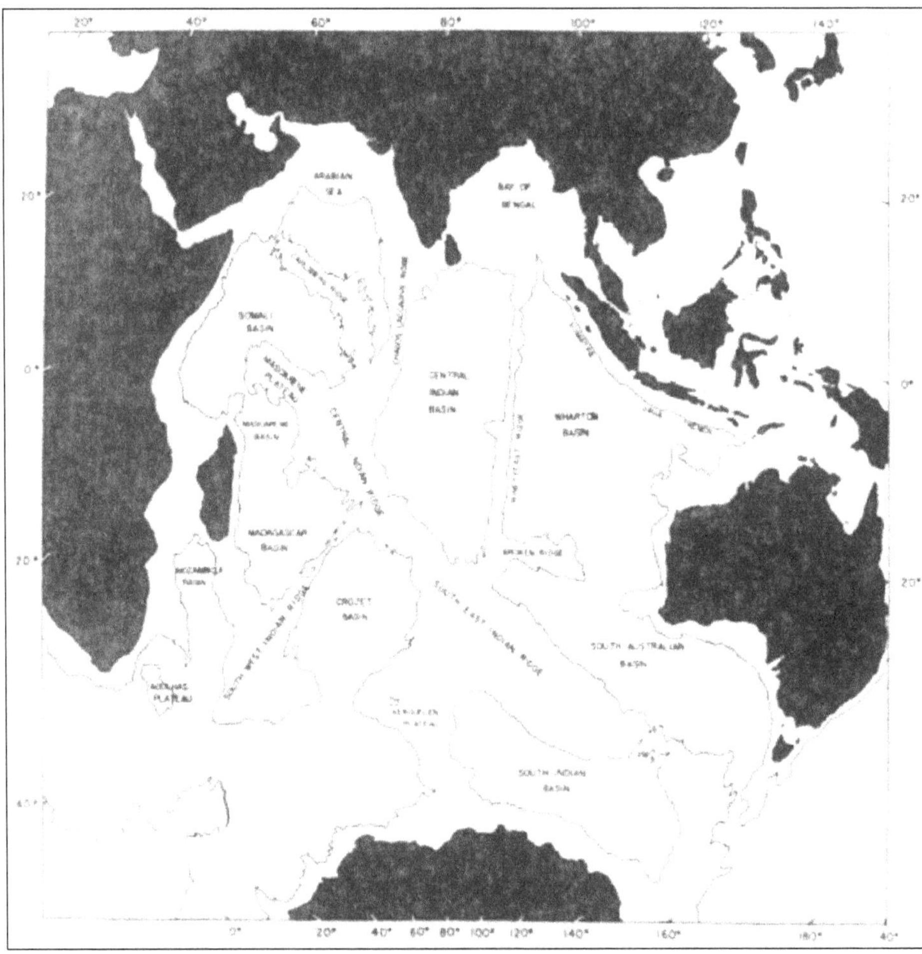

Figure 15.3: Bathymetric Features of the Indian Ocean (Fraser, 1986).

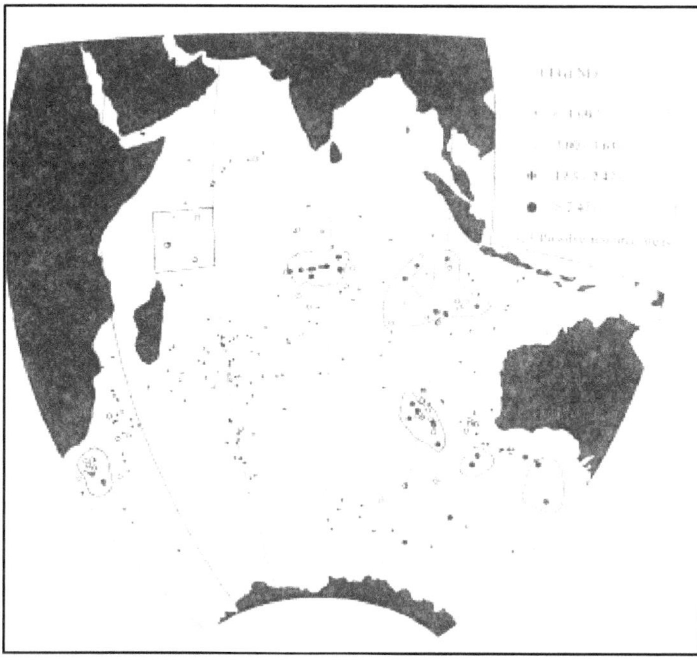

Figure 15.4: Distribution of Nodule Grade in the Indian Ocean.

Nations expert meeting). The Indian ocean is the most complex among the three major oceans (Sclater, 1977), as shown in Figure 15.3.

Since the manganese deposits are present over a wide variety of isolated environments, they are extremely varied in composition. The distribution of nodule grade throughout the Indian Ocean is as follows:

The distribution of the nodule grade (Ni +Cu + Co) in the Indian Ocean varied significantly (Figure 15.4). For instance, the Central Indian Basin and the South Australian Basin have a grade ≥ 2.47 per cent whereas within the EEZ of Mauritius the grade is quite low, < 1 per cent.

The manganese distribution in the Indian Ocean is as follows (Figure 15.5):

It can be noted that in the Central Indian Basin Wharton Basin as well as in the South Australian Basin manganese nodules of paramarginal grade (≥ 18 per cent); nickel (Figure 15.6) and copper (Figure 15.7) tend to follow manganese. Cobalt (Figure 15.8) exist in lowest concentrations, on average between 0.15 to 0.3 per cent, except in seamount nodules, where the latter tend to exist in unusually high amounts and are associated with high cobalt and very low copper concentrations. Also, cobalt (^{3} 0.45 per cent) is higher in the western half of the ocean, near the Seychelles islands.

Around the EEZ of Mauritius manganese content for most of the samples is between 10 to 20 per cent whereas nickel content varied between 0.2 to 0.6 per cent,

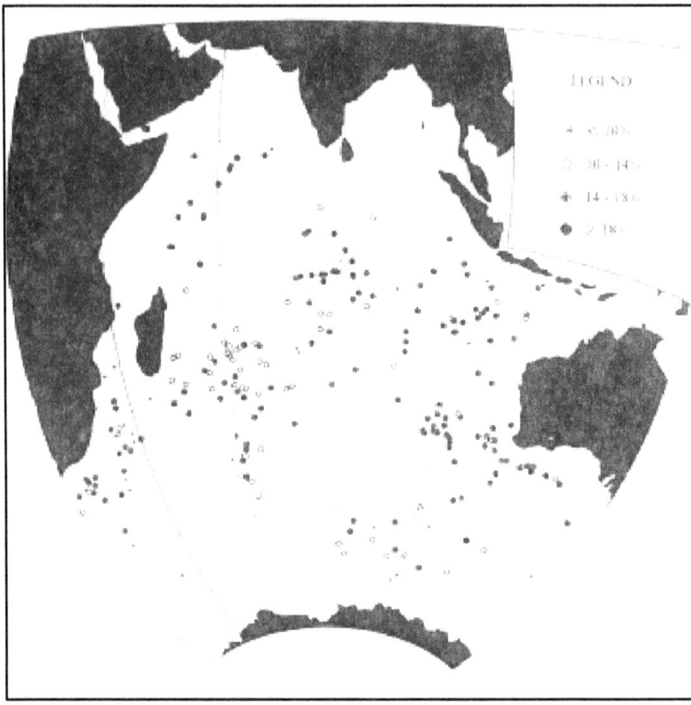

Figure 15.5: Manganese Distribution in the Indian Ocean.

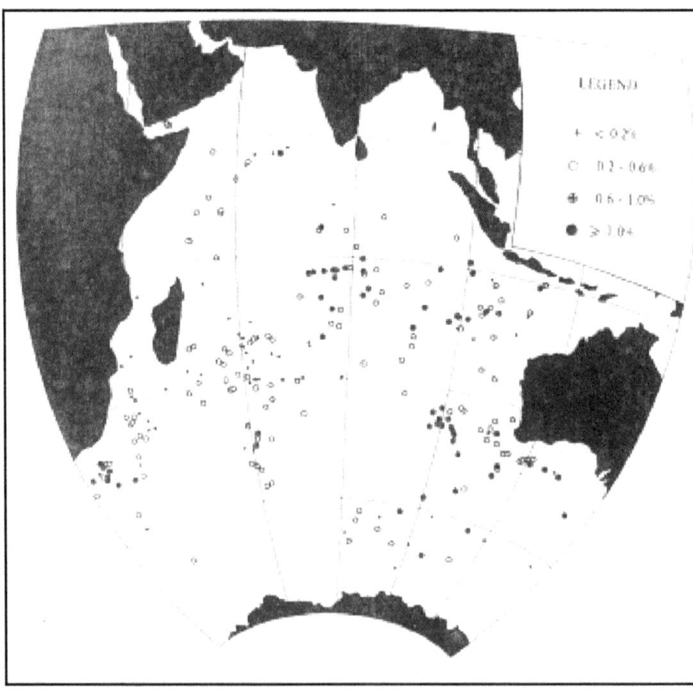

Figure 15.6: Nickel Distribution in the Indian Ocean.

Figure 15.7: Copper Distribution in the Indian Ocean.

Figure 15.8: Cobalt Distribution in the Indian Ocean.

and that for copper < 0.2 per cent. However, cobalt is found to exist in higher amounts ($\geq$ 0.45 per cent).

The five areas identified where manganese nodule deposits are likely to be of any economic interest were:

Seychelles Region

The Seychelles are an island group in the north end of the Mascarene Plateau. Surrounding these island groups are seafloor manganese deposits. The manganese deposits consist of both nodules and crust. Analysis of about 80 per cent of the samples revealed that they were of crust nature, leading to the conclusion that the crust is the prevalent form of manganese deposit in this region. However the grades of the nodules and crusts, expressed in terms of total nickel, copper and cobalt content, are too low to be economically viable. On the other hand, the nickel equivalent corresponding to a paramarginal grade is 1.98 per cent, and that corresponding to the minimum resource grade of 1.32 per cent. Cobalt has been identified (> 0.4 per cent) to the west of the Mauritius-Seychelles Ridge extending from the Mascarene Basin to the west of Seychelles (Somali Basin), passing through Tromelin and Agalega. Part of this cobalt-rich area is within the EEZ of Mauritius (Fogoonee, 1985). More data on the nodule abundance need to be collected.

Central Indian Basin

It lies between the Chagos Laccadive Ridge on the west and the Ninetyeast Ridge on the east. At a depth of 5000 to 5500 m nodule rich in nickel and copper were found. Those nodules are of submarginal and paramarginal grade and extend over an area of 700,000 km^2. The average grade is 2.49 per cent above that required for first-generation mining. Analysis showed that nickel and copper ^{3} 1.5 per cent. Copper content is unusually high in relation to nickel. These data showed that nodules occur somewhat more frequently south of 10° S than north of that latitude, perhaps because of higher sedimentation rates (3-10 mm/1000 years) to the north (Hayes, 1970). More explorations are needed to establish the true value for these deposits.

Wharton Basin

It is bordered to the west and south by Ninetyeast Ridge and Broken Ridge, respectively. It is deeper that the Central Indian Ocean (5500-6000 m) and sedimentation rate is slightly higher. The nodules recovered are not considered as paramarginal resources since most of the samples have a grade of 1.6 per cent but the average is only 1.39 per cent. Cu/Ni ratios are highly variable, but the mean is about the same as that for typical areas in the Clarion-Clipperton zone. Although nodule deposits in some parts of this region may be abundant for mining, the grade is probably too low to make these deposits a resource.

South Australian Basin

This Basin is of considerate economic potential since it consists of an extensive deposit of abundant nodules with a mean grade of 2.09 per cent. The south Australian Basin is strongly influenced by the Antarctic Bottom water, which has eroded the South East Indian Ridge between 120-130° E. Due to this erosion activity abundant deposits of manganese nodules have developed just south of the ridge in the Tasman Basin, and in the western sector of the South Australian Basin (Kennett, 1976). The latter is named "Southeast Indian Ocean Manganese pavement" and is described as "characteristically covered by dense packings of manganese nodules to form a manganese pavement". On the whole, nodule deposits in the South Australian Basin are of low grade and are far from meeting the criteria for paramarginal resources. However, because they are abundant and close to Australia, they may be eventually mined more profitably that other deposits with somewhat higher grades.

Aghulas Plateau

It is situated off the southern tip of Africa (Barrett, 1977). The nodules have nickel, copper and cobalt content almost similar to those in the Seychelles region. The mean nickel-equivalent grade, however, is only 1.32 per cent, minimal for resources.

Exploitation of these nodules has not been successfully achieved up to now due to some controversial aspects related to manganese nodule mining and they are:

Distribution

The distribution of nodules is governed by a number of factors. Most importantly, there is need for low rates of sedimentation of land-derived material and organic remains from the surface from the surface layer of oceans. In addition, the local environmental factors such as bottom current activity, can lead to erosion and redistribution of sediments. There should also be presence of nuclei suitable for ferromanganese oxide precipitation to be initiated which in turn will determine the nodule distribution. Most nodule nuclei are composed of fragments of volcanic rock and these nodules tend to be abundant in submarine volcanic areas.

Compositional Variability

Manganese nodules differ in composition throughout the oceans. Regional variations in the abundance of the economically important elements namely nickel, copper, cobalt and manganese have also been noted. For instance, nodules from seamounts tend to be richer in cobalt, but have low concentrations of nickel and copper; those from the mid-Ocean ridges are rich in iron, and those from abyssal seafloor are rich in nickel and copper but are low in cobalt. Since nickel and copper are of greatest economic interests, the nodules from the abyssal seafloor prove to be of highest potential economic value. However, at depth > 3500 m, the nodule composition varies considerably. Large areas of the deep ocean floor contain nodules with < 1 per cent each of nickel and copper and thus are unlikely to be of economic interest. The minimum value required for deposits to be commercially exploited, nickel and copper should exceed 2-3 per cent combined. This usually occurs in a few widely scattered areas namely parts of northern tropical Pacific, south Pacific and the Central Indian Ocean.

Origin of Ore Grade Nodules

To predict the best spots for exploration, it is vital to have a full understanding of the origin of ore grade nodules. It is believed that submarine hydrothermal activity is important in the production of ore grade nodule (Raab, 1972). However, it was postulated that the metals supplied by hydrothermal processes are largely precipitated very close to the vents and are strongly fractionated from each other over short distances on the seafloor (Cronan, 1978). Owing to this fact, it is unlikely that hydrothermal activity would cause uniform enrichment of elements such as nickel and copper in manganese nodules over wide areas of the ocean floor as in the case of the northeastern equatorial Pacific. A more plausible mechanism involves the selective extraction of metals from sea water. It is noted that the nodules from the northeastern equatorial Pacific and the Central Indian Ocean have certain common features. Firstly, they occur on the margins of the equatorial zone where large-scale sinking of organic remains occur due to high biological activity. Secondly, they are associated with siliceous ooze sediments below the depth at which organisms composed of calcium carbonate dissolve. Thirdly, they contain todorokite, one of the two principal minerals in the nodules, as their principal mineral phase and which has a structure suitable to accommodate enriched nickel and copper. Such nodules may form as a result of the extraction of metals from the surface waters by organisms and their transport to the seafloor where they are liberated into the bottom waters and interstitial waters of the sediments as the organisms dissolve. The metals are then available to take part in chemical reactions in the uppermost layers of sediments, leading to the formation of manganese nodules.

Economic Potential

The potential economic value of the deep sea manganese nodules is one of the most controversial aspects of these deposits (Mero, 1965). Though they contain sufficient supplies of may metals to last for thousands of years based on the rates of consumption now-a-days, only a fraction contain economically mineable concentrations of the metals of interest. There are several problems in calculating reserves of manganese nodules; for instance, there is no universal agreement as to what grade and abundance of nodules actually constitutes a mine site. Many workers have assumed a minimum abundance of 10 kg/m^2 to be required to support a mining operation. However, this value was considered to be too high (Bastien-Thiry, 1977). Similarly there is no firm agreement on the average and cut-off grades of nickel and copper necessary to support a mining operation, although the 2.27 per cent and 1.8 per cent respectively for nickel and copper proposed by J. T. Kildow *et al.*, is gaining acceptance (Kildow, 1976).

Estimates of the number of possible mine sites must take into account potential problems in actually mining the deposits, such as the degree of ruggedness of the seafloor.

Conclusion

The five areas identifies is characterized by a different type of manganese nodule deposit. The most promising site for mining is in the Central Indian Basin. Up to date, data collected for manganese nodules in the Mauritian territorial waters do not make the latter economically viable. However it has to be noted that these nodules have higher amounts of cobalt compared to the other areas identified. More explorations are to be conducted to have a better perspective of the presence of the nodules.

Concerning metalliferrrous muds, the latter are more promising since they occur in less deep waters and important metals such as zinc, copper and gold can be recovered near segments of the Ridge closest to Rodrigues. Studies about deep sea minerals near the ridges, fracture zones and ECS need to be done in view of investigating deposits of polymetallic sediments. It will depend on the stock value of the base metals and raw materials and someday Mauritius will be at a key position in the collection and processing of ore from the deep-sea - whether in its own EEZ or from nearby permits granted by the ISA to other countries (Personal Communication, Dyment, 2012).

References

1. Barrett, D.M, 1977. Agulhas Plateau off southern Africa: A geophysical study. *GSA Bulletin*, 88(6), pp. 749.
2. Bastien-Thiry, H., 1977. *Energy and Mining Journal*, 178(86).
3. Cronan, D.S., 1978. Manganese nodules: controversy upon controversy. Endeavour, 2(2), pp. 82.
4. Frazer, J. Z., 1979. The reliability of available data on element concentrations in seafloor manganese nodules. In: Deepsea mining, MIT press.
5. Hayes, J.D., 1970. Geological investigations of the North Pacific. GSA Memoir, 126, pp. 83-119.
6. Kennett, J.P. and Watkins, N.D., 1976. Regional deep-sea dynamic processes recorded by late Cenozoic sediments of the southeastern Indian Ocean. *GSA Bulletin*, 87, pp. 321-329.
7. Mcro, J.L., 1978. Marine mining I, 243.
8. *Marine Technology Society Journal* 1982 Special Issue Polymetallic sulphides, 16(3).
9. Yales, J., Spagni, D. and Kcanc, J., 1986. Endeavour, 10(1).

References from Books

10. Cameron, H.M., 1983. Indigenous Raw materials for Industry. Royal Society London, pp. 57-61.
11. Mero, J.L., 1965. The Mineral resources of the sea. Elsevier, Amsterdam.
12. Raab, W. and Mayland, M., 1972. Marine manganese deposits. Elsevier, Amsterdam.

References from Other Literature

13. Dyment, J. (2001). Deep-Sea Exploration of Central Indian Ridge at 19°S. Pre-cruise Report.
14. Fagoonee, I., 1985. Prospects for marine-based industries for Mauritius. In: Proceedings of Natural Conference on Industrialisation, University of Mauritius, Mauritius.
15. Frazer, J.Z. and Wilson, L.L., 1979 Manganese nodule deposits in the Indian Ocean SIO reference series.

16. Holser, A.F., 1976. Manganese nodule Resources and mine site availability. Report to US department of the Interior.

17. Kildow, J.T., Bever, M.B., Dar V.K. and Capstaff A.E., 1976. Assessment of Economic and Regulatory conditions affecting ocean minerals Resource development. Report to US dept of interior, MIT.

18. Mustafa, Z., 1984. Offshore Mineral Resources 2nd International GERMINAL Seminar. In: Brest Proceedings, pp. 509-540.

19. Sclater, J.G. and Heirtzler, J.R., 1977. An introduction to Deep sea Drilling in the Indian Ocean. In: Ocean Geology and Biostratigraphy Am. Geophys. Union.

20. United Nations, 1982. Seabed Minerals Series, 1, London, pp. 7.

21. Phosphorite nodules: Fertilisers from the Sea, 1984. PREST. Marine Technology Brief no. 5.

22. Proceedings of United Nations expert meeting on seabed mineral resource assessment,1977.

23. Tyler, S., 2003. An investigation of evidence for Ridge-Hotspot Interaction along the Central Ocean Ridge. Report for SPC-EU Deep Sea Minerals, Southampton Oceanography centre, Southampton, UK.

24. Mauritius Oceanography website displaying the brochure to sensitise people on both national and international scale.

25. http://moi.gov.mu/publication/CLCSBrochure.pdf (accessed 2 Sept. 2012)

26. Wikepedia webpage showing definition for abyssal plain.

27. http://en.wikipedia.org/wiki/Abyssal_plain (accessed 3 Sept. 2012)

Chapter 16

Geological Resource Potential of Vietnam and Current Mining Practices: An Overview

Tran Thanh Hai

Hanoi University of Mining and Geology, Hanoi, Vietnam
E-mail: tranthanhhai.humg@gmail.com

ABSTRACT

The territory of Vietnam occupies the eastern margin of the Indochina peninsular. Current geological features of the country were the consequence of successive assembly and collision between the Indochina, South China and other plates during the Phanerozoic to form present-days Southeast Asia. The unique geological setting with long and complex tectonic history has created favorable environments for the generation of rich geological resources including minerals, water, energy and geological heritages. Mineral resources comprise many types of metallic, non-metallic and industrial minerals with thousands of deposits and occurrences have been discovered, some with significant potential. Metallic resources consist of ferrous, base, precious and rare metals. None metallic minerals comprise dominantly gemstones whereas industrial minerals compose a wide variety of materials. The exploration and mining activities in Vietnam have a long history of development. However, the mineral sector has not sustainably developed, due to the shortage of advanced knowledge, skill and expertise in research, exploration, extraction, processing and beneficiation of minerals as well as inefficient management. Systematic adoption of modern technology in along with improvement of knowledge and expertise in mineral industry from basic research to exploration, extraction and beneficiation, as well as management system are very important for the sustainable development of mineral industry.

Keywords: *Vietnam, Indochina, Geology, Mineral resources, Geological resources, Mining industry.*

Introduction

The Socialist Republic of Vietnam, with a land area of more than 320 thousands km^2 and more than 1 million km^2 of sea area and its continental shelf/sea bed, occupies the eastern side of the Indochina Peninsula and western part of the East Sea, forming a relatively narrow S-shaped band of landmass that extends nearly 2000 km along the eastern margin of Indochina Peninsular (Figure 16.1).

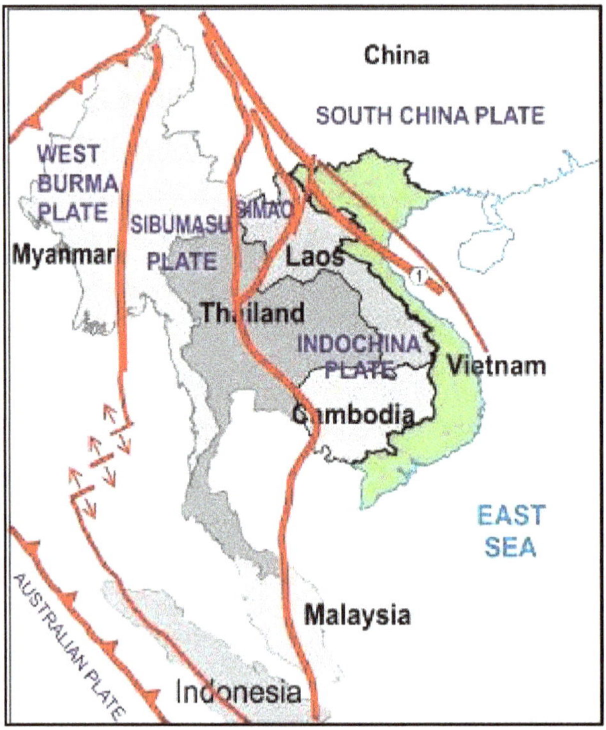

Figure 16.1: Schematic Diagram Showing the Geographic and Tectonic Position of Vietnam Relative to Southeast Asia.

Modified from Metcalfe, 2005 and other sources. Thick red line (1) is the Song Ma fault zone, which has been interpreted to represent a paleo-suture zone.

Large part of Vietnam's landmass is covered by mountainous terrains whereas the lowland parts occur as deltaic plains of major river systems or as narrow coastal plains. The country is well endowed with many types of geological resources, which are scattered both inland and under the sea bed. The present mineral production includes various types of minerals and energy resources. However, in comparison with the country's mineral wealth, the mining sector is underdeveloped. This paper presents a brief overview of the country's geology, its major mineral resources and their potential, as well as an outline of the current statute of mining practices.

General Geological Feature of Vietnam

Tectonic Setting

Recent regional geological works have mostly supported that the current geological setting and features of Vietnam were the direct results of assembly and collision between the Indochina and South China plates (*e.g.*, Hutchison, 1989; Metcalfe, 2005; Tri and Khuc, 2011). Most of workers favour that the assembly between the two plates has taken place along the Song Ma suture zone (Figure 16.1). The collision has consequently resulted in the fold-thust deformation and destroyed many of the primary features of lithotectonic assemblages. In addition, current structural frame of Vietnam's territory was also strongly affected by numerous plate-plate collisional events and docking of several plates to Eurasia during the Mesozoic to Early Cenozoic to form present-days Southeast Asia (Figures 16.1 and 16.2). Cenozoic formation of the East Sea has been interpreted as the results of southeast-ward terrane extrusion along major intercontinental faults that caused by crust indentation during India-Eurasia collision during Early Cenozoic (*e.g.*, Tapoinnier *et al.*, 1990; Metcalfe, 2005; Hall, 2009).

General Geology

The territory of Vietnam constitutes complicated geological features. The country is underlain by multiple reactivated Paleozoic to Mesozoic fold belts overlaying Precambrian cratonic basement inliers that comprises variably deformed and metamorphosed, lower to high-grade metamorphic rocks of various compositions (Figure 16.2; *e.g.*, Luong and Bao, 1988; Thanh and Khuc, 2006; Tri and Khuc, 2011). Large part of the country is covered by thick Paleozoic siliciclastic and carbonate sedimentary successions (Figure 16.2), which unconformably overlay the Precambrian complexes. These rocks are partly unconformably overlain by Late Paleozoic carbonate and terrigeneous deposits of shallowing upward marine environments. All above successions were either partly intruded by intrusive bodies or unconformably overlain by volcanogenic and terrigeneous rocks of Latest Paleozoic to Early Mesozoic ages.Middle Mesozoic to Early Cenozoic supracrustal rocks are restricted and comprised intermontane molasse deposits and lesser volcanic assemblages. Cenozoic sedimentary sequences infilled the fault-controlled intracontinental basins, deltas as well as cover the East Sea basin floor. Whereas Late Neogene-Early Quaternary continental flood basalts cover large part of Central Highland and southeastern part of the territory.

The pre-Cenozoic volcano-sedimentary assemblages were intruded variably by igneous rocks of various types and ages (Figure 16.2; Luong and Bao, 1988; Tri and Khuc, 2011). In the north and central part of the country, the intrusive rocks generally form small bodies that are scattered along generally northwest-southeast trend belts, where as in the southeastern part, magmatic complexes are more abundant and form north-south to northeast-southwest trending belts (Figure 16.2). Magma composition ranges from ultramafic, mafic, intermediate, and felsic varieties (see Tri and Khuc, 2011), in which ultramafic and mafic intrusives commonly occur as small bodies, many of them are fault-controlled and have been interpreted as remnants of ophiolite

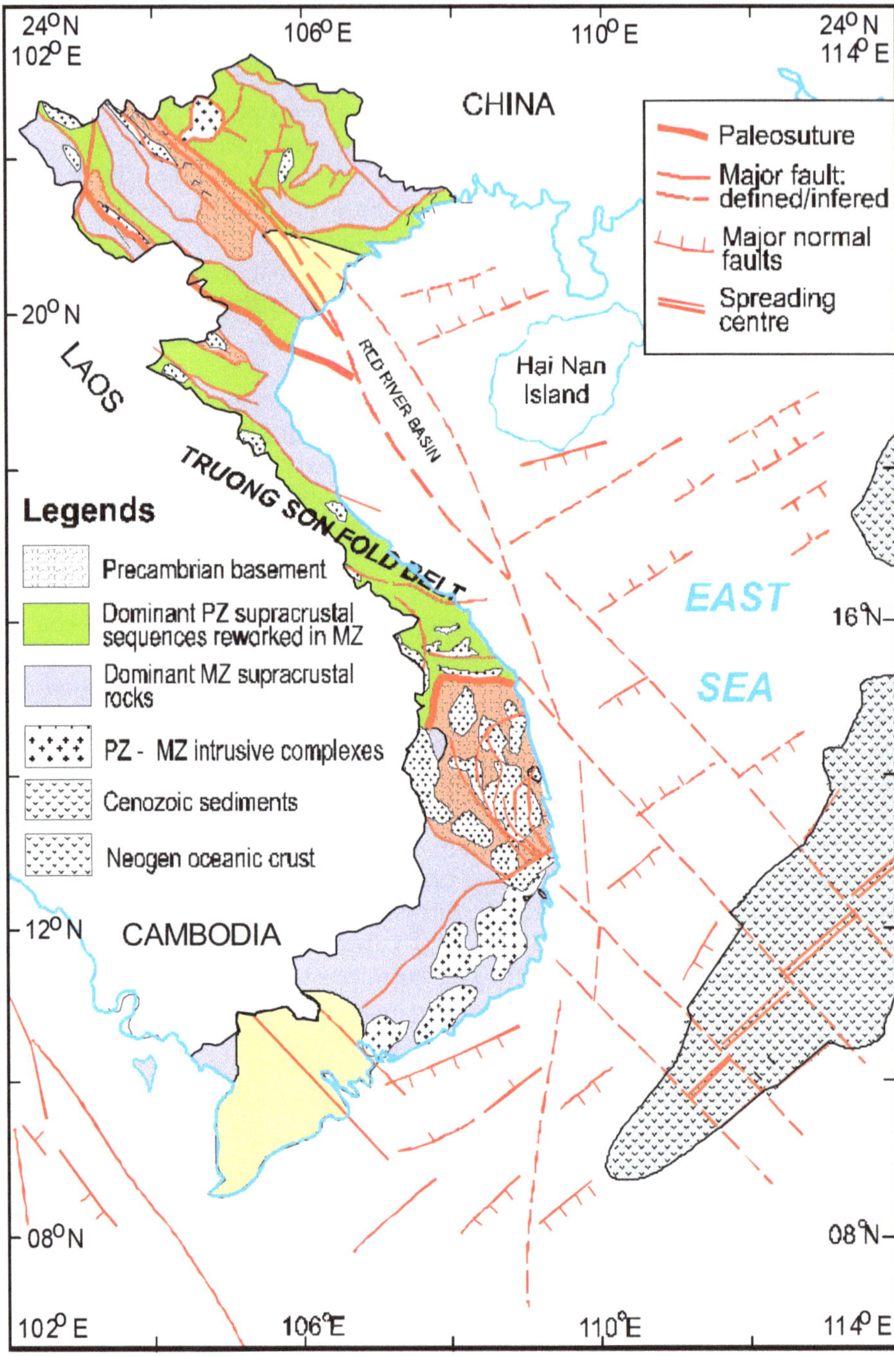

Figure 16.2: Generalized Litho-tectonic Map of Vietnam Showing the Distribution and Relationship of Major Litho-tectonic Elements of the Landmass Part of the Country. Modified from Luong and Bao (1982); Tri (1977, 1979); Tri and Khuc (2011) and other sources.

sequences (Tran *et al.*, 2011; Tri and Khuc, 2011). Felsic intrusives, on the other hands, commonly form large batholith complexes, many of them have been considered to derived from S-type, orogeny generated magmas. The magma generation was governed by major tectonic events such as magmatic arc, continental rifting, ophiolite obduction or orogenic process, which took place mostly in Paleozoic to Early Cenozoic (*e.g.*, Hutchison, 2007; Metcalfe, 2005; Lepvrier *et al.*, 2008; Tri and Khuc, 2011 and references therein).

All above rocks were variably deformed by successive orogenic events that were originated by multiple-phase plate collision and terrane assembly that extended from Paleozoic to Early Cenozoic to form present-days Southeast Asia. The most important orogenic event in Indochina area has been recognized so far is the collision between South China and Indochina plates, which led to the 'Indosinian' orogeny event (Fromaget, 1937, 1941) and formed Truong Son fold belt in central Indochina (Figure 16.2). Many geological workers have argued that Indosinian orogeny took place during early Mesozoic (*e.g.*, Fromaget, 1937, 1941; Hutchison, 1989; Lepvrier *et al.*, 2008 and references therein). However, it is growing evidences that a much earlier collisional event that took place during Early to Middle Paleozoic had shaped Indochina and the Indosinian event in fact were consequence of Sibumasu-Indochina plate collision during Early Mesozoic (*e.g.*, Metcalfe, 2005; Carter *et al.*, 2001; Carter and Clift, 2008 and references therein). Although most of orogenic events has ceased in Early Cenozoic, the ongoing intra-continental tectonic indentation caused by India-Eurasia plate collision during Cenozoic had led to the fragmentation and partly rifting of the proto-Southeast Asia landmass, which led to the formation of the East Sea and numerous interacontinental fracture zones (*e.g.*, Tappoinier *et al.*, 1990; Hall, 2009). Presently, some major intra-continental structure zones within the country are remained active, which frequently produce earthquake and the other geological hazards.

Mineral Resources of Vietnam and Current Mining Practices

General Features of Geological Resources of Vietnam

The unique geological setting with long and complex tectonic evolution of Vietnam's territory has created favorable environments for the generation of diverse natural resources, some with significant potential. Currently published data has recorded thousands of deposits and occurrences of many types of geological resources including mineral, energy, ground water, and geological heritages (*e.g.*, Tri, 2000; Tri and Khuc, 2011).

Mineral occurrences, which occur scatteredly in entire territory (Figure 16.3), comprise many types of metallic, non-metallic and industrial minerals.

Energy resources comprise coals, petroleum, and uranium (Tri and Khuc, 2011). Most of the coals are in the form of anthracite, which hosted by Late Triassic molasses deposits or as lignite in Neogene sediments, which mostly occurred in northern Vietnam. Petroleum resources include oil and gases are mostly hosted by Cenozoic sedimentary rocks and related structures in off-shore basins (Figure 16.3).

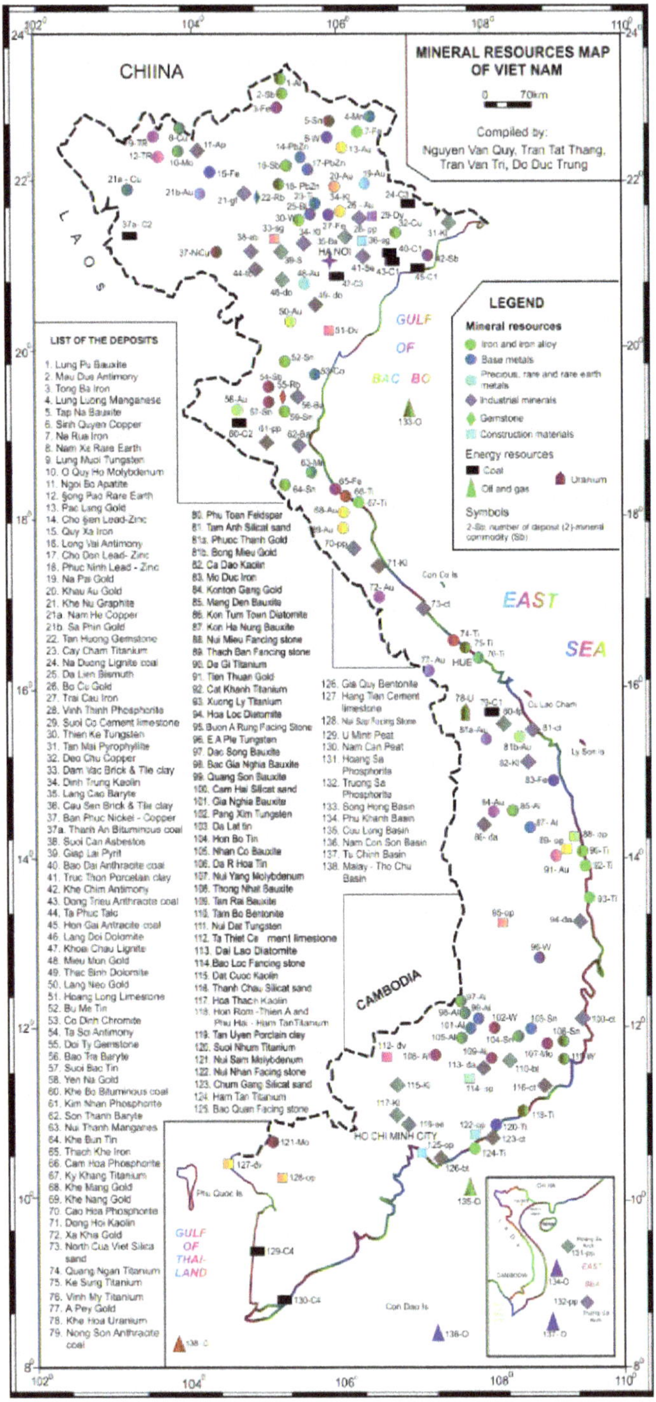

Figure 16.3: Generalized Map of Mineral and Energy Resources of Vietnam (Modified after Tri and Khuc, 2011).

Ground water resources in Vietnam are very rich, hosted in many types of rocks and many controlled by regional structure features. The ground water resources comprise 3 types: fresh, thermal and mineral water, in which nearly 100 major occurrences of mineral water has been recorded (Tri and Khuc, 2011).

Geological heritages are also major types of geological resource of Vietnam, characterized by unique landscapes, geosites and geotopes of outstanding universal values. Two areas in Vietnam been awarded the World Natural Heritages including Ha Long Bay and Phong Nha- Ke Bang, one area is World Biosphere Reserve (Cat Ba Archipelago), and one Global Geopart (Dong Van Karst Plateau) by UUNESCO.

Mineral Resource Potential

Metallic Minerals

Ferrous Minerals

Ferrous metals comprise predominant iron, manganese, chrome, titanium minerals. Iron ore are abundant and current data has recorded 216 occurrences, which distribute mainly in the north and central regions (Figure 16.3; Tri and Khuc, 2011). The iron sources derived from many origins including magmatic, skarn, hydrothermal and metasomatic, sedimentary, metamorphic and weathering. A number of the occurrences have been explored and the iron resource/reverse has been proved. However, many other showings and occurrences have not been adequately evaluated.

Significant chromite deposit has been recorded so far is the Co Dinh Chromite deposit located in north central Vietnam (Figure 16.3). The chrome is associated with are serpentinite body. The chromite is enriched in the placers that derived from weathering and reworking of the serpentinite.

Titanium in Vietnam is abundant and scatteredly occurred in northern and central part of the territory. Titanium mostly occurs in the form of ilmenite, which are enriched in the coastal placers to form large deposits (Figure 16.3) or disseminated within some ultramafic-mafic intrusive complexes.

Base Metals

Non-ferrous base metals in Vietnam comprise copper, lead-zinc, tungsten, tin, antimony, aluminum, zircon and other elements such as silver, cadmium, indium, etc. Many occurrences and deposits of copper have been discovered in Vietnam but proved resource of this type of mineral is limited (Tri and Khuc, 2011). Major copper mineralized district occurred in northwest of Vietnam, where copper and associate minerals are hosted by metasomatized Precambrian metasedimentary rocks. Other copper occurrences occur together with nickel either in ultramafic-mafic complexes, in mafic volcanic rocks or as part of hydrothermal veins.

Lead-zinc minerals also abundant in northern and central part of Vietnam with numerous small-scale occurrences and showing have been discovered (Tri, 2000; Tri and Khuc, 2011). The lead-zinc mineralization comprises two major types: carbonate-hosted and magmatic hosted. Many of the occurrences/showings, however, have not been investigated systematically and their origin and potential are largely unknown.

Major wolfram occurrence locates in Thai Nguyen province, northern Vietnam, where wolfram is hosted by skarn- and greisen -style polymetallic mineralizations together with Cu, Au, Bi, and Fluorite with a total reserve and resources of 110 million tones (Tri and Khuc, 2011). Other small showings/occurrences including hydrothermal wolfram- or wolfram-cassiterite quartz veins that are closely associated intrusive rocks.

Bauxite ore is major source for aluminum in Vietnam, which is abundant and mostly associated with the lateritic-bauxitic weathering profile of Neogene-Quaternary basalts occurring in Central Highland with an inferred resource of up to 1.6 billion tones. Other smaller occurrences of bauxite are hosted by Late Permian carbonate rocks that occurred in northernmost of Vietnam.

Other nonferrous minerals in Vietnam have been discovered including numerous showings of tins, bismuth, molybdene and other metals (Figure 16.3). However, current geological works have not taken place sufficiently to define the potential and significance of these minerals.

Precious and Rare Metals

Gold is typical source of precious metals in Vietnam and abundantly occurred with hundreds of showings and deposits (Tri and Khuc, 2011) have been discovered and mined for thousands of year. Gold occurrences comprise several types, including both primary and placer gold. Primary gold occurs abundantly in variable types of rocks in northern and central Vietnam. The gold either controlled by shear-zone hosted quartz veins within various types of sedimentary or volcanic-hosted quartz veins. Placer gold, on the other hand, commonly occurs within alluvial deposits occurred along the river systems. Except for few major showings and deposits in northern and central of Vietnam have been explored systematically, most of gold occurrences in Vietnam have not sufficiently evaluated.

Other sources of precious metals in Vietnam include silver, platinum, tantalo-niobium, beryllium, lithium, and indium have been recognized in Vietnam. However, the nature of the mineral occurrences and their significance is remained to be solved.

Rare Earth Element (REE) group is the major form of rare metals. Main occurrences of this group of minerals occur in northwestern Vietnam including 4 major deposits of both HREE and LREE (Figure 16.3). The mineralization comprise both weathering and primary categories that hosted by various rock types. REE has also been discovered in other areas. However, their origin and potential are still the subject for further studies.

Non-Metallic Minerals

Ruby and sapphire as well as other gemstones are widely discovered in Vietnam, including numerous showings in northern, north-central and in Central Highland (Figure 16.3; Tri and Khuc, 2011). However, their potential is largely unknown due to the lack of systematic investigation and in-depth study. Current data show that the predominant sources of none-metallic mineral resources in Vietnam are ruby and sapphire. Both ruby and sapphire are generally hosted by either metamorphosed carbonate rocks (skarn-type) or high-alkaline basalts.

Industrial Minerals

Vietnam hosts very rich sources of industrial minerals of numerous types, some of which are of great potentials. Typical of industrial mineral sources include apatite, phosphorite, barite, fluorite, pyrite, serpentine, peat, porcelain clay, dolomite, white limestone, magnesite, graphite, feldspar, muscovite, beryl, and silica sand and many other minerals of small occurrence (Tri, 2000; Tri and Khuc, 2011; Figure 3). Some of the industrial minerals such as apatite, fluorite, white limestone have been proved to be of significant potential.

Construction Materials

Vietnam hosts great potential and variety of construction materials (Tri and Khuc, 2011; Figure 16.3). Many types of material such as clay, limestone, facing and construction stone, sand, etc... are enormous and available in most part of the country.

Current Mining Practice in Vietnam

Mining Practice

Exploration and mining activities have been carried out in Vietnam at least since the Bronze Age for bronze, gold, silver, and some other non-metallic metals as well as construction materials. The mining activities were evolved along with the development of feudal societies. However, the exploitation of the country mineral resources was not intensified until and after the French occupation of Indochina in late 18th Century, when many types of minerals were systematically explored, mined and exported to France (eg., coal, gold, tin, chromite, zinc, antimony, apatite, talc and some other minerals). During the first half of 20th Century before 1945, mining activities in Vietnam were intensified by the French and many mining districts had been established. After the regaining of independence from France (in 1954), a vigorous exploration and the restoration of mining industry was carried out in Northern Vietnam whereas such activities in Southern Vietnam were not available until after the re-unification of the country in 1975.

The exploration activities during 1960s-1990s have led to the discoveries of many important metallic and non-metallic deposits, as well as coal, oil and gas fields and as such have created the basis for the mining industry to develop. Since the 1980s, exploration and mining activities have been intensified to meet the demand of materials for rapid economic growth of the country, which has led to the formation of natural resources-based economy. At present, major energy and mineral production includes oil, natural gas, coal, cement, chromite, iron ores, gold, and many other minerals for domestic use as well as for export. Several foreign companies are involved in exploration for minerals and mining activities.

Problems Related to Mining Activities

Despite rapid growth in mining industries, the minerals sector in Vietnam is largely underdeveloped and many available mineral reserves remain poorly evaluated and unexploited. Except a limited number of large deposits such as anthracite and lignite, copper, gold that is being controlled by the state-owned or foreign-owned mining companies, which applied advanced mining practices, most of small-scale mining activities are conducted by small mining companies and individuals with uncontrolled artisan mining methods. The application of modern technology to develop a highly valued and/or export-orientated mineral industry is largely not available due to the shortage of advanced knowledge, skill and expertise in research, exploration, extraction to processing and beneficiation as well as the weakness of management policies. Environmental protection is largely ignored in many mining areas. As consequence, air, ground and water pollution as well as uncontrolled mine wastes have created many environmental problems. Such shortcomings have prevented the mineral industry of Vietnam to develop sustainably and produce many negative environmental impacts, which could lead to unsustainable development of the country.

Concluding Remarks

The unique geological features have created a wide range of geological resources in the territory of Vietnam. Many type of minerals, ranging from metallic, none metallic, industrial minerals are present. Some of them such as apatite, bauxites, rare earths, wolfram are of international significant. Other mineral such as gold, gems, copper, chromite, titanium and other minerals are commercially viable. These mineral resources are hosted by unique litho-tectonic entities that formed by a long-lasted tectonic.

Mineral resources of Vietnam has been explored and exploited and contributed significantly to the national economy. However, the mining sector of Vietnam still underdeveloped. Many types of mineral resources are still inadequately evaluated and unexploited. The small-scale development with out-of-date mining techniques and ineffective managing practice have created environmental largely underdeveloped and many available mineral problems, which may threaten the sustainable development of the mineral sector.

Systematic application of modern technology in mineral industry along with improvement of knowledge, expertise in mineral sector from basic research to exploration, extraction and processing and beneficiation, as well as creation of sound management policies are all important for the sustainable development of mineral industry. In addition, increasing involvement of the high-tech foreign companies in mineral sector will also accelerate the development of highly valued and profitable export-orientated mineral industry, improve environmental protection and secure the sustainability of the mining sector as well as national economy.

References

1. Carter, A., Roques, D., Bristow, C. and Kinny, P., 2001. Understanding Mesozoic accretion in Southeast Asia: Significance of Triassic thermotectonism (Indosinian orogeny) in Vietnam. *Geology* 29, 211–214.
2. Carter, A. and Clift, P.D., 2008. Was the Indosinian orogeny a Triassic mountain building or a thermotectonic reactivation event? *C. R. Geoscience* v. 340, pp. 83–93.

3. Fromaget J., 1937. Carte géologique de I'Indochine au 2.000.000. *Serv. Géol. Indoch.*, Hanoi.

4. Fromaget J., 1941. L'Indochine francaise, sa structure géologique, ses roches, ses mines et leur relation posible avec la tectonique. Bull. SGI, XXVI/2. Hanoi, 140 p.

5. Hai Thanh Tran, Ngo Xuan Thanh, Jacqueline Halpin, Khin Zaw, 2011. The occrrence of ophiolite-style assemblages along Sino-Vietnam border, Northeastern Vietnam and its implication to the tectonic evolution of Northeastrn Indochina. *Proceedings of Inter. Conf. on Geology of Indochina* (GEOINDO 2011), Khon Kaen, Thailand, pp. 479-488.

6. Hall, R., 2009. The Eurasian SE Asian margin as a modern example of an accretionary orogen. Geological Society, London, Special Publications v. 318, pp. 351-372.

7. Hutchison, C.S., 2007. Geological Evolution of South-east Asia, second edition. Geological Society of Malaysia, Kuala Lumpur, 433 pp.

8. Lepvrier, C., Nguyen Van Vuong, Maluski, H., Phan Truong Thi, Tich Van Vu, 2008.Indosinian tectonics in Vietnam, *C. R. Geoscience*, v. 340, pp. 94– 111.

9. Metcalfe I., 2005. South-east Asia. In: Elsevier Encyclopedia of Geology, Elsevier Ltd.

10. Tapponnier, P., Lacassin, R., and Leloup, P.H.·Schärer, U., Zhong, D., Wu, H., Liu, X., Ji. S., Zhang, L., and Zhong, J., 1990. The Ailao Shan/Red River metamorphic belt: Tertiary left-lateral shear between Indochina and South China. Nature 343, 431–437.

11. Tong-Dzuy Thanh, Vu Khuc (eds). 2006. Stratigraphic Units of Vietnam. Vietnam National University Publishing House, Hanoi. 526 p.

12. Tran Duc Luong, Nguyen Xuan Bao (eds), 1988. Geological map of Vietnam scale 1:500.000 scale. General Department of Geology and Minerals of Vietnam, Hanoi.

13. Tran Van Trt (ed.), 1977. Geology of Northern Vietnam. Science and Technics Publishing House, Hanoi, 355 p. (In Vietnamese).

14. Tran Van Tri, 1979. Geotectonic evolution of Northern Vietnam. XIV Pacific Sci. Congress, Khabarovsk. Com. B, Sect. II, p. 48-49.

15. Tran Van Tri (ed), 2000. Mineral Resources of Vietnam. Department of Geology and Minerals of Vietnam, Hanoi, 214 p. (In Vietnamese).

16. Tran Van Tri and Vu Khuc (eds.), 2011. Geology and Earth Resources of Vietnam. Publishing House for Science and Technology, Hanoi, 645 p.

Chapter 17

Zimbabwe Proposes to Host a Pan-African Minerals University of Science and Technology (PAMUST)

Francis P. Gudyanga

Ministry of Science and Technology Development
Livingstone Building, 48 Samora Machel Avenue
P.O. Box 2265, Harare, Zimbabwe
E-mail: fpgudyanga@zarnet.ac.zw

ABSTRACT

At a meeting of the World Bank with African leaders in 2001 it was suggested to the World Bank that world class African Institutes of Science and Technology (AISTs) in sub-Saharan Africa are perhaps Africa's greatest need.

The AISTs are now being developed as a pan-African university system aimed at developing the highest level of technical learning facilities to the continent of Africa. Their focus is on what is missing in our developmental system *i.e.* post-graduate education, research and innovation linked to entrepreneurship. The ultimate vision is to develop four AIST campuses across Sub-Sahara Africa (SSA)—one each in the north, south, east, and west of the continent.

In January 2005 the African Heads of State and Government agreed to establish the four AISTs as follows: Western Africa: Abuja (Nigeria) – focussing on Energy and Petrochemical Engineering; Eastern Africa: Arusha (Tanzania) – with a niche area of Life Sciences (Bio-engineering and Agriculture); and Northern Africa: Ouagadougou (Burkina Faso) focussing on Water and Environmental Engineering. The fourth Institute in Southern Africa was mandated to have a thrust on Mineral beneficiation and value addition. Zimbabwe proposes to establish the pan-African post-graduate university for training and research in mineral beneficiation and value addition for the African continent.

Zimbabwe is centrally located in a mining region that comprises Central and Southern Africa. The country has a robust combination of institutional support systems for the mining industry that includes the Ministry of Mines and Mining Development, Chamber of Mines, Educational and Research institutions and Mining Associations.

The Pan African Minerals Institute of Science and Technology, PAMUST, is proposed to be housed at the Scientific and Industrial Research and Development Centre (SIRDC) in Harare. SIRDC has enough land, modern research and development laboratories, a Library and documentation centre, Internet and Telecommunications networks, Cafeteria and Conference facilities.

Keywords: *AIST, Postgraduate, Mineral beneficiation, Value addition, PAMUST.*

Introduction

Even though Africa is indeed now free from conventional colonialism, poverty, hunger, diseases and civil wars are still a commonplace. Despite its enormous resources of nature, Sub Sahara Africa (SSA) in particular has not been able to adequately harness its endowments for its sustainable development. Sub-Sahara Africa's failure to harness its resources of nature is attributable to lack of the required critical mass of human capital with appropriate knowledge and skills (value-added human capital), technology, infrastructure for innovation and entrepreneurship, and incentives to innovate.

In February 2001 IMF Managing Director, Horst Köhler, and World Bank President, James Wolfensohn, travelled to Africa to meet with African leaders as a follow-up to their commitment at the Prague 2000 Annual Meetings. During one of the meetings, the President of South Africa, Nelson Mandela lamented the knowledge deficit in Africa to Mr. Wolfensohn. President Mandela went on to suggest to the World Bank that world class African Institutes of Science and Technology (AISTs) in sub-Saharan Africa were perhaps Africa's greatest need.

The African Institutes of Science and Technology (AIST) have now been conceived and developed as a pan-African university system aimed at developing the highest level of technical learning facilities to the continent of Africa. Their focus is on what is missing in our developmental system *i.e.* post-graduate education, research and innovation linked to entrepreneurship. The ultimate vision is to develop four AIST campuses across Sub-Sahara Africa (SSA) - one each in the north, south, east, and west of the continent. The AISTs are being developed into a network of world-class institutions, to promote excellence in the sciences within the continent, through strong linkages with existing national universities, and research institutions. AISTs would draw on the decentralized model that has been successfully implemented by the Indian Institutes of Technology (IITs). The concept was further pursued by the World Bank in collaboration with some Africans both on the continent and in the diaspora. The initiative represents a global effort to foster Sub-Saharan Africa's economic growth and development through the promotion of excellence in science and engineering and their applications. The AISTs are tasked to train and development the next generation of African scientists, engineers and technologists, who will impact on the continent's development through the application of science, engineering and technology.

Objectives of the AISTs

- To catalyze the development of world class science and technology in Africa in ways that can stimulate economic growth and generate employment, and turn brain drain into brain gain;
- To provide the best facilities for post-graduate and post-doctoral studies and research;
- To bridge between research and industrial development through strong link with industry for technology transfer; and
- The establishment of technology parks around the AISTs for seeding innovations, with technology-cum-business incubators for spin-off technological firms

Establishment of the Institutes of Science and Technology

During the 4th Ordinary Session of the African Union Assembly held in Abuja in January 2005 the African Heads of State and Government agreed to establish 4 AISTs - one each in the Western, Eastern, Northern, and Southern Africa. The AIST project is being run by the Nelson Mandela Institution supported by the World Bank, among other development partners. NMI's Board of Directors is chaired by the Dr. Ngozi Okonjo-Iweala, the Finance Minister of the Federal Republic of Nigeria. The NMI Scientific Board is chaired by Phillip Griffiths from the Institute for Advanced Study at Princeton University in the USA. The AISTs are being established as follows:

Western Africa: Abuja (Nigeria)

The focus of the African University of Science and Technology (AUST) in Abuja is on *Energy and Petrochemical Engineering*. The institution was established in 2007 as the first Pan-African Network of Institutes of Science and Technology and Centres of Excellence located across the continent. The AUST in Abuja is able to count on substantial pool of diasporan scientists and engineers who are teaching and conducting research in the best universities across the world. Strong partnerships have already been established with the African Institute of Mathematical Sciences (AIMS) in Cape Town and with Indian Institute of Technology Bombay (IIT). It is being funded by the World Bank Group, amongst other development institutions.

Eastern Africa: Arusha (Tanzania)

The niche area of the Nelson Mandela African Institute of Science and Technology (NM AIST) is Life sciences and Bio-Engineering taking advantage of the immense bio-diversity in the East Africa and SSA countries. As an initial move to accept the initiative, the Government of Tanzania offered 3,285 acres of land in Karangai area in Arusha. The Government has also provided about 1,394 hectares of land on which the NM AIST-Arusha is being developed, plus the start-up finances for the rehabilitation and face-lifting of existing buildings and infrastructure (hitherto owned by a national institution which has been relocated elsewhere), and some new constructions. To-date the commitment by the Government of Tanzania amounts to USD 56.9 million. The university started enrolling its Masters and Ph.D. students in 2011.

Northern Africa: Ougadougou (Burkina Faso)

The International Institute for Water and Environmental Engineering (2iE), located in Ouagadougou (Burkina Faso), aims to provide training in Water Engineering and Environment. Water, the environment, energy and infrastructure engineering are the priority areas in a development and integration policy that applies to the whole of Africa. The goal of 2iE is to train a sufficient number of men and women in the skills needed to meet this development challenge. The business plan was developed in 2006. The goal of the International Institute for Water and Environmental Engineering is to become a centre of excellence on the African continent, recognized throughout the world for its training and research in the areas of water, energy and the environment.

Southern Africa

The outstanding campus to be established in Southern Africa is to focus on *Mineral beneficiation and value addition.* Over the years when the AISTs were being conceived it was always strongly felt that Southern Africa should host the campus that will have a thrust in mineral beneficiation and value addition. Independent of the above AIST initiative SADC commissioned studies 1995 and 2001 to assess mineral processing facilities in the region with a view to rationalizing their uses. Southern Africa is therefore expected to have a centre that will train a cohort of African minerals professionals for the continent.

Rationale for a Continental Institute on Minerals

The African mineral industry has evolved largely as a producer for foreign export markets. The developed countries are focusing primarily on technological development programmes that emphasise value-added and high margin products, while encouraging developing countries (mostly African) to remain primary producers of minerals. Value addition of mineral resources cannot be achieved by labour intensive methods alone due to the cost of labour, process inefficiency and low productivity levels in the African region when compared to developed countries, but rather via a technology-based approach.

One-way to stretch Africa's resources is to recycle on the continent. There is definitely scope to investigate and develop technologies for incorporating recycling streams in our extractive metallurgy processes. For example, recycling of PGM containing auto catalytic converters with PGM extraction, or recycling lead acid batteries with lead extraction, etc. These integrated technologies may not be important for Europe due to their limited mining activities.

While Africa may have some R&D facilities in the primary and secondary processing stages there isn't a single well-functioning R&D facility that offers post-graduate training comparable to leading schools of mines in Europe (Royal School Mines, established in 1851), North America (Colorado School of Mines, established in 1873) or Australasia (Western Australia School of Mines, established in 1929). Africa needs such a world-class R&D institution with linkages to the manufacturing sectors. Southern Africa may be considered the natural home for such and institution within the AIST pan-African university system due to the diversity and large deposits of minerals found in the region. Such a centre would rival the leading schools of mines (RSM, CSM, WASM) where many Africans are currently trained at the continent's immense costs.

The university is envisaged to have emphasis on the following areas:

- Geosciences
- Mining Engineering
- Extractive Metallurgy
- Materials Science and Engineering
- Mineral Business Studies

The Institute will have the following characteristics:

- Postgraduate programmes (Masters, Ph.D. and post-doctoral.)
- Pan African in student and staff composition
- World class institution with the most advanced facilities
- Large enrollment numbers

Benefits to Africa by Establishing a Continental Institute on Minerals

The Institute will attract leading world experts, local and diasporan Africans to research on African minerals with a view to adding value to them with the creation of opportunities for downstream industries. With post-graduate dissertations and theses being carried out at the Institute, Africa will become a rich repository and a minefield of information on minerals to which the continent will have a privileged, immediate and unrestricted access. The Institute will be uniquely positioned to leverage research funding to Africa for training scientists and engineers as part of a continental strategy on manpower development for other institutions in the continent. The institute will contribute immensely in the transformation of Africa from a resource-based economy to a knowledge-based economy as a result of minerals value addition, value added products, downstream industries, business opportunities, employment creation, skills enhancement, and technology transfer.

Proposal to Establish the Minerals Institute in Zimbawe

Zimbabwe proposes to host the pan-African post-graduate university dedicated to the teaching and training of high calibre mineral professionals in mineral beneficiation and value addition for the African continent.

About Zimbabwe

Zimbabwe is centrally located in a mining region that comprises Central and Southern Africa. It is well connected regionally by road, rail, air, telecommunication and power grid. Zimbabwe is a landlocked country in Southern Africa sharing a border with the Republic of South Africa in the south, and is bounded on the southwest and west by Botswana, on the north by Zambia, and on the northeast and east by Mozambique.

The capital of Zimbabwe is Harare and the official languages are English, Shona and iSindebele. It is sparsely populated with about 13 million people or 34 inhabitants per square km. The country enjoys a pleasant temperate climate with 365 days of sunshine.

Following the establishment of the Inclusive Government in February 2009, Zimbabwe has managed to emerge out of a crisis, that way establishing a stable macroeconomic environment in support of sustainable growth. Since 2009, inflation has been contained within single digit levels, averaging 3 per cent up to date. These gains were supported by robust policies centrering around cash budgeting for attaining fiscal prudence, introduction of multiple hard currencies which contained inflation, broad liberalization measures which removed a multitude of controls as well as a host of other market based measures in support of productive sectors and infrastructure rehabilitation and development.

Consequently, the country has since been enjoying spectacular economic performance with GDP growth rates of 5.4 per cent in 2009, 9.6 per cent in 2010, an estimated 9.4 per cent in 2011and a projected 5.6 per cent for 2012. The immediate impact of this positive performance has been renewed investors' interest and confidence in doing business in Zimbabwe. Investment opportunities abound in agriculture, mining, agro-processing, infrastructure, tourism and manufacturing among others. The relatively strong and sound banking sector with about twenty five banks stands ready to support investors. Mining, in particular offers immense opportunities with vast reserves of platinum, gold, coal, chrome, uranium, tin, copper, asbestos, and the recently discovered large reserves of diamonds.

Zimbabwe is an active member of the African Union (AU), the Southern Africa Development Community (SADC), and the Common Market for Eastern and Southern Africa (COMESA). The country enjoys cordial relations with African countries many of whom are represented by diplomatic missions in the country. It is a member of the United Nations and the Non-Aligned Movement (NAM) and a member of the World Bank and the IMF. Zimbabwe hosts several UN agencies such as the UNDP, UNESCO, WHO, UNICEF, UNHCR, etc. Zimbabwe is signatory to regional legal instruments that seek to promote the development of mining on the continent. It signed and ratified the SADC Protocol on Mining on 8 September 1997 and 19 July 1999, respectively. It is a founding member of the Kimberly Process System (KPS) for diamonds. While there are no COMESA protocols on mining, Zimbabwe is signatory to the COMESA Treaty that provides the mandate to member states to cooperate in the development and management of natural resources, environment and wildlife. Zimbabwe is also a signatory to the SADC Protocol on Science, Technology and Innovation whose

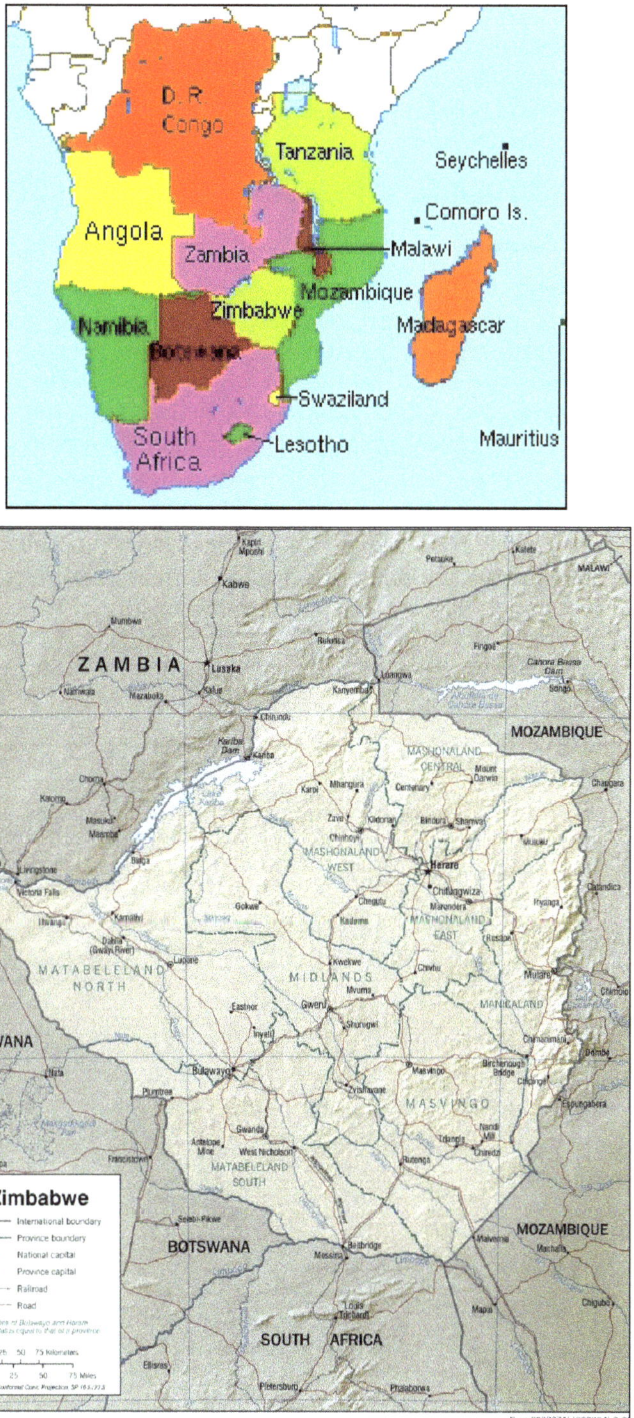

Figure 17.1: Maps of SADC and Zimbabwe.

objective is to foster co-operation and promote the development, transfer and mastery of science, technology and innovation in Member States through the establishment of mechanisms for regional collaboration, harmonisation of science, technology and innovation policies and promotion of gender equity in the teaching of basic science at all levels of educational system.

The Basis for Zimbabwe's Proposal

Vibrant Mining Sector

Zimbabwe has a vibrant mining sector with over 60 minerals 40 of which being mined. The mineral resources are dominated by two prominent geological features: the famous Great Dyke and the ancient Greenstone Belts, also known as Gold Belts. The Great Dyke is a spectacular body containing the world's largest high grade chromite resources base, second largest resource of platinum group metals as well as significant reserves of copper and nickel. The minerals found in Zimbabwe include the following; gold, coal, platinum group of metals, copper, nickel, iron, and diamonds. It is the leading or second leading producer of minerals such as gold, platinum group metals, chrome, coal, asbestos, lithium and diamonds. It is also host to several other minerals found nowhere else in the world such as rare earths with very great potential for further discoveries. Mine development in the history of the country has been responsible for the growth of settlements which have grown to be major cities and towns. The development of rail infrastructure occurred to service the mines and mining communities.

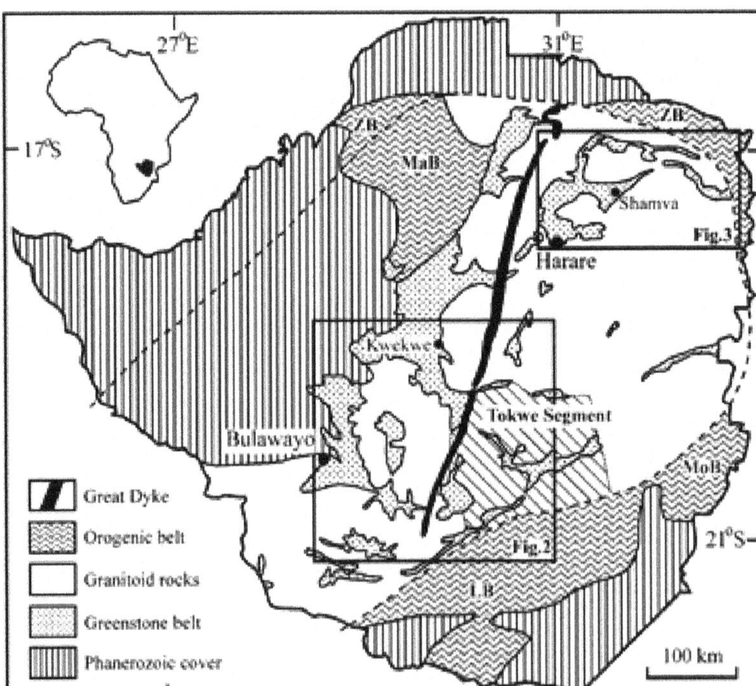

Figure 17.2: Geological Map of Zimbabwe.

Sound Educational System in Support of the Mining Sector

As of 2010 Zimbabwe, according to the UNDP Digest, has had an adult literacy rate of 92 per cent which is the highest in Africa. It has a robust education system that starts with 2 years of pre-school, 7 years of primary and 6 years of secondary schooling before students can enter university in the country or abroad. In addition to universities (presently 14 in number) there are several Polytechnics, Technical and Teacher Training Colleges. Zimbabwe's trained mining professionals are sought after world-wide.

Institutional Support for the Mining Industry

The mining sector has a robust support system that includes the Ministry of Mines and Mining Development, Chamber of Mines, Educational and Research institutions and Mining Associations.

The Ministry has the following support systems: The Department of Mining Promotion and Development which is responsible for policy formulation, facilitating and promoting investment for sustainable mining and value addition of the country's mineral resources for the socio-economic wellbeing of all Zimbabweans. The Mining Law Administration Department works towards creating a conducive environment through granting and preservation of mining titles timeously and regulating mining activities. The Zimbabwe Geological Survey, established in 1910, provides quality geo-scientific information and services for the exploration and development of the country's mineral resources. The Geological Survey compiles the geo-scientific information of the crust through the systematic geological mapping of rock formations and general cataloguing and description of mineral resources of the country. The Mining Engineering Department regulates the mining industry to ensure sustainable mining practices and management for the purpose of creating a safe and healthy mining environment. The Metallurgy Department works towards ensuring sustainable and responsible mineral processing, verifying the technical and economic feasibility for beneficiation, marketing of mineral resources and mineral production for the benefit of all Zimbabweans.

The Chamber of Mines is a private sector voluntary organization established in 1939. The members include mining companies, suppliers of machinery, spare parts and chemicals, service providers including banks, insurance companies, consulting engineers, and various mining related professional bodies and individuals. The mining company members of the Chamber produce about 90 per cent of Zimbabwe's total mining output. The Chamber of Mines has been involved in a number of social corporate responsibilities.

Educational and Research Institutions in Support of the Mining Sector

The University of Zimbabwe established in 1957 has the following Departments that support and train for the Mining Industry: Geology, Mining, Metallurgy, Chemistry, Geo-physics, Institute of Environmental Studies, Surveying and Geo-Informatics. Other departments that play a supporting role of providing technology include Electrical Engineering, Mechanical Engineering, Civil Engineering, and Computer Science. The Institute of Mining Research (IMR) established in 1969 and located at the University of Zimbabwe is jointly managed by the University of Zimbabwe and the Ministry of Mines and Mining Development. The Ministry of Mines and Mining Development provide funding by way of an annual grant to IMR which is supplemented by income from contract research and externally funded research projects. Most of the research undertaken is applied in nature, but theses for higher degrees have been produced in applied geochemistry, applied mineralogy, economic geology, metallurgy, rock mechanics and occupational health and safety. It offers mining related consultancy services and facilities.

The Zimbabwe School of Mines, managed by the Ministry of Mines and Mining Development, is a dynamic regional institution that serves the Mining Industry in Zimbabwe and the SADC region by providing competency-based training for high calibre, hands-on technicians. Training that is provided in Mining, Geology, Mine Survey, Mineral Processing and Extractive Metallurgy, Metallurgical Assaying is relevant to the mining industry and is constantly revised to meet the modern technological requirements.

The National University of Science and Technology has the following Departments that support and train for the Mining Industry: Chemical Engineering offering courses in Minerals Engineering, Quantity Surveying, Applied Chemistry, Applied Physics, and Environmental Science and Health.

The Scientific and Industrial Research and Development Centre (SIRDC) is Zimbabwe's Technology Centre established to carry out specialised research in strategic areas of the country's industry. Its Institutes that conduct research activities linked to the mining industry include: Environmental Sciences, Metallurgical Research, Centre for Analytical and Testing Services, Production Engineering, Geo-Information and Remote Sensing, Electronics and Communications, National Metrology.

Ancillary Facilities

The Mining industry is served by a well maintained system of tarred roads, railway links and an efficient electricity grid. There are numerous industries that manufacture a wide range of mining inputs. Hospitals, schools, hotels, and recreational facilities service the mining communities. In addition Zimbabwe has a well developed and modern banking system with a relatively vibrant stock exchange. The Mining Museum in Kwekwe and the National Archives contain volumes of information on mining in Zimbabwe. The National Social Security Authority (NSSA) provides services for occupational health to the mining workforce.

Proposed Location of the Pan-African Minerals University of Science and Technology (PAMUST)

PAMUST in Harare

The Pan-African Minerals University of Science and Technology will be located in Harare which has features and amenities that come with the biggest metropolis in Zimbabwe and as the Capital City. It has the highest concentration of professionals who can support and be supported by this institution. Harare has the following universities with which PAMUST will share facilities and experiences: University of Zimbabwe with 10 faculties including Faculties of Science, Engineering; Harare Institute of Technology; Harare Polytechnic; Headquarters and largest Campus of the Zimbabwe Open University; the Catholic University in Zimbabwe; the Women University in Africa; The Research Council of Zimbabwe, the Zimbabwe Council for Higher Education, the Zimbabwe Academy of Sciences, the Zimbabwe Institute of Engineers, the Zimbabwe Association of Inventors; other professional organisations are located in Harare. It has the Headquarters of virtually all the Mining Houses and Companies operating in Zimbabwe; the Chamber of Mines is located in Harare; Harare is the home to most of the big manufacturing industries;

Harare is easily accessible by road, rail and air with several flights in and out of Harare International Airport per day. Harare is the Seat of Government where policy and decision makers can be promptly engaged.

PAMUST at SIRDC

PAMUST will be housed at the SCIENTIFIC AND INDUSTRIAL RESEARCH and DEVELOPMENT CENTRE (SIRDC). The Scientific and Industrial Research and Development Centre (SIRDC) is Zimbabwe's Technology Centre established by the Government of Zimbabwe in 1993 under the provisions of the Research Act of 1986 with the mandate to carry out industrial research. SIRDC is comprised of 12 specialized research institutes and employs scientists and engineers with expertise in the following broad areas:

- ✰ Biotechnology
- ✰ Building Technology
- ✰ Cleaner Production Technology
- ✰ Electronics and Communications
- ✰ Energy Technology
- ✰ Environmental Science
- ✰ Food and Biomedical Technology
- ✰ Geographic Information Systems and Remote Sensing
- ✰ Information and Communication Technology
- ✰ Metallurgy
- ✰ Metrology
- ✰ Production Engineering
- ✰ Analytical Testing Services
- ✰ Polymer Science

SIRDC conducts scientific research leading to product development and technology transfer to Zimbabwe and regional stakeholders. The Centre also offers consultancy services and training in the above areas of specialization. SIRDC's mission is "To provide Zimbabwe and the region with technological solutions for sustainable development" while its vision is "To be the leading Centre for the development of Zimbabwe and the region through reduction to practice of technologically developed products and processes".

The infrastructure at SIRDC includes:

- ✰ Modern research and development laboratories
- ✰ Library and documentation centre
- ✰ Internet and Telecommunications networks
- ✰ Cafeteria and Conference facilities
- ✰ Space for expansion and field work

SIRDC Institutes Linked to Mining Activities

The following institutes at SIRDC conduct activities that are directly linked to the mining and mineral processing industry.

Environmental Sciences Institute (ESI)

The Environmental Sciences Institute (ESI) provides technological services to national and regional stakeholders in order to enhance environmental performance of industrial products and services at local and international markets. ESI hosts the National Cleaner Production Centre (NCPC) of Zimbabwe. ESI's portfolio of activities includes the following:

- ✰ Re-vegetation of old tailings dumps.
- ✰ Design and construction of landfills for general waste disposal.
- ✰ Design and construction of tailings dams.
- ✰ Engineered covers for final closure of tailings dams and rock waste dumps.
- ✰ Capacity building or training in any of the above.
- ✰ Water quality monitoring and contaminated soils remediation.
- ✰ Environmental Impact Assessments for all mining and development projects.
- ✰ Training of contractors in all components of occupational, health, safety and environmental issues.
- ✰ Ambient air quality monitoring on mining sites, boiler stack measurements, fumes measurements and dust measurements.
- ✰ Land suitability classification.
- ✰ Natural resources management (soil, water, minerals, forests and biota)
- ✰ Ecological research
- ✰ Industrial Ecology
- ✰ Environmental Management Systems (EMS)
- ✰ Resource Efficiency and Cleaner production (RECP) and Sustainable Consumption
- ✰ Effluent analysis and management
- ✰ Water treatment

- Wastewater treatment
- Resource recovery, reuse and recycling
- Wetland restoration and bioremediation
- Solid waste management
- Hazardous waste management
- Integrated waste management
- State of environment reporting
- Waste water analysis
- Contaminated soils analysis

Metallurgical Research Institute (MRI)

The Metallurgical Research Institute (MRI) undertakes focused research and development in metallurgical, materials, foundry and other forming technologies to develop and commercialize a portfolio of bankable projects in the specialized area of metallurgy such as:

- Materials and Processes
- Physical Metallurgy
- Foundry Technology
- Extractive Metallurgy

The MRI focuses on:

- Development and provision of elution, electro winning and carbon regeneration plants and custom services.
- Treatment of refractory gold ores and dumps.
- Development, production and supply of mining, mineral processing, automotive and agricultural equipment and spares.
- Mining, processing and refining of gold and platinum group metals (PGMs).
- Value addition of local ores.
- Development and production of refractories.
- Providing technical consultancy and training services to industry.
- SGI and ADI manufacturing technology.
- Mineral extraction and refining (gold, chromium, Platinum Group Metals, Diamond, etc.).
- Process plant flow design.
- Mineral value addition, *e.g.* tantalite.
- Process improvement and optimisation, *e.g.* optimisation of slag/alloy separation for a high carbon ferrochrome producer.

Centre for Analytical and Testing Services

Centre for Analytical and Testing Services is equipped with state of the art analytical and testing equipment including equipment for geochemical analysis of:

- Base metals
- Heavy metals
- Platinum Group Metals (PMGs)

The array of equipment includes the following:

- Inductively Coupled Plasma Optical Emission Spectrometer (ICP OES) – for chemical, geo-chemical, and water/wastewater analyses
- Atomic Absorption Spectrometer (AAS) – for metal analysis
- Fourier Transmission Infra-Red Spectrometer (FTIR)
- Gas Chromatograph- Mass Spectrometer (GC-MS)
- Analytical and precision balances
- Elemental Analyser (C, N, S)

Production Engineering Institute (PEI)

PEI develops competitive engineering products, machinery and manufacturing systems for the socio-economic development. PEI's areas of focus include the following:

- New Materials: namely, Polymers, Composites, Ceramics
- Production Design Technologies
- Manufacturing Process Design and Improvement
- Machinery Design and Development
- Maintenance and Reliability
- Automation and mechanisation of production lines
- Project management
- Development and implementation of ISO 9000 quality systems.
- Maintenance Engineering and Management (namely, Preventive, Predictive, Reliability Centred)
- Design, evaluation and planning of plant layouts
- Manufacturing systems design
- Establishment and implementation of best manufacturing models and customised management systems for different manufacturing organisations.
- Training of industrial personnel on all aspects of world-class manufacturing to ensure sustainability of implemented manufacturing solutions and continuous improvement
- Capacity studies using Industrial Engineering Techniques
- Logistics management.

- Air conditioning and ventilation systems design (HVAC)
- Design and development of material handling systems
- Productivity Improvement
- Vibration Analysis
- Auto CAD

Geo-Information and Remote Sensing Institute

The Geo-Information and Remote Sensing Institute (GRSI) contributes to the achievement of sustainable environmental resource utilization in Zimbabwe and the region. GRSI provides spatial data analysis solutions that result in the management of the environment and natural resources in a more sustainable manner than would be possible without considering the spatial aspects. The emphasis is on the use of remote sensing (RS) techniques and geographic information systems (GIS). GRSI has expertise in the following areas:

- Remote Sensing Applications
- Creation and Management of GIS Database
- Mineral Exploration
- Early warning of natural disaster
- Earth and Water Resources Management
- Satellite image processing and classification
- Dust Fallout Modelling
- Digitising hardcopy maps for use in a GIS

Electronics and Communications Institute (ECI)

The work programme for ECI includes the following:

- Design and development of original electronics and communication equipment for industrial instrumentation and control systems, process monitoring and automation
- Design and development of original or client specified embedded and intelligent micro-controller based applications
- Technology transfer and training services in specialized areas of electronics and communications such as micro-controller technology, circuit design, electronic instrumentation and control
- Repair and maintenance services for a wide range of industrial electronic and communication equipment and instruments
- Consultancy and support services for industrial automation, control and communications

National Metrology Institute (NMI)

The National Metrology Institute (NMI) is the Custodian of National Measurement Standards. NMI provides traceable measurements and calibration services to the public and private sectors of the economy. These services enable industries to compete on the global markets as well as meet statutory and contractual obligations. NMI's focus areas include:

- Maintenance of National Measuring Standards
- Equipment Calibration Services
- Developing New Measuring Standards

Land

SIRDC has enough land to house PAMUST which is expected to be a separate university campus. SIRDC has 160 hectares and currently occupying only half of it. The SIRDC campus is contiguous to state land that can be given to the new institution for expansion and field work.

Post-Graduate University

It is planned that PAMUST will be a separate and primarily a post-graduate university with most of the work there being that of research. This fits in well with the core business of SIRDC. There will be mutual benefits to both PAMUST and SIRDC as they share and exchange research facilities and experiences.

Zimbabwe's Commitments in Support of the Proposal to Host PAMUST

Financial Commitment

Zimbabwe is committed to provide financial support in the form of seed funding, annual support as per university's allocation, and the establishment of an endowment fund.

Commitment to Excellence

Zimbabwe is also committed to excellence in the provision of higher education through the Zimbabwe Council for Higher Education (ZIMCHE). The Zimbabwe Council for Higher Education (ZIMCHE) was created through an Act of Parliament (Chapter 25:27) with a mandate to promote and coordinate education provided by institutions of higher education and to act as a regulator in the determination and maintenance of standards of teaching, examinations, academic qualifications and research in institutions of higher education. It is further mandated to design and recommend an institutional quality assurance system for higher education, that is, a system whereby the courses, programmes and degrees offered by institutions are evaluated on a regular and objective basis, and to recommend to the Minister responsible for Higher and Tertiary Education institutional quality assurance standards. ZIMCHE will therefore be the gatekeeper of standards in higher education institutions such as PAMUST.

Research excellence is also assured by the Research Council of Zimbabwe. The Research Council of Zimbabwe (RCZ) is a statutory body established in terms of the Research Act of 1986. It is mandated to promote, direct, supervise and co-ordinate research. RCZ also provides an exceptional forum for interaction and discussion for the mutual benefit of Government, academia and industrialists. In its endeavour to promote research and publicise Zimbabwean research at a global scale, the RCZ maintains a research database that informs on research developments in Zimbabwe and beyond its borders. The purpose of the database is to provide an integrated and comprehensive collection of data covering all public domain research in the country. The Research Council of Zimbabwe gave birth in 2004 to a nascent but active Zimbabwe Academy of Sciences (ZAS) - a learned society - whose objective "is the pursuit, encouragement and maintenance of excellence in the fields of science and technology in order to promote the advancement of and practice of science and technology for the development of the nation and the benefit of mankind". Fellows of the Academy are elected by virtue of their respective achievements in the field of science and technology and are regarded as being of exceptional merit and distinction and who can be expected to significantly assist the Academy in achieving its objectives. PAMUST students and staff will therefore have ample opportunities to interact with the most eminent scientists in Zimbabwe.

Commitment to the Pan-African Nature of the Initiative

Zimbabwe has been supportive of this pan-African initiative of AIST university system since its conception in 2001. Zimbabwe is an active member of the African Union (AU), the Southern Africa Development Community (SADC), and the Common Market for Eastern and Southern Africa (COMESA). The country enjoys cordial relations with African countries many of whom are represented by diplomatic missions in the country. It is a signatory to various regional and continental protocols.

Zimbabwe is committed to the Pan-African nature of the proposed institution in terms of the composition of staff and students. The Government of Zimbabwe will guarantee the issuance of visas and work permits for African students and staff who satisfy the country's immigration requirements and are eligible for enrolment and employment at the Institute.

Commitment and Ability to Attract/Create Leadership and Faculty of Highest Quality

Zimbabwe is committed to sound recruitment strategies for the new Institution on the basis of merit for (i) part-time, permanent, semi-permanent faculty, (ii) faculty development including promotion from junior to senior, tenure framework and standards. Zimbabwe has run its institutions of higher learning on this basis and has consistently attracted and retained high quality staff from both within Zimbabwe and abroad. Zimbabwe will endeavour to recruit the best staff for the Institution from across the continent.

Connection to Industry

The mining and manufacturing industry will be open to student and staff for field work. The Institution will encourage initiatives for promoting spin-offs to new industries. This PAMUST initiative has the support of the mining sector represented by the Chamber of Mines and the Ministry of Mine and Mining Development.

Outreach

As with other institutions of Higher Education the PAMUST will be encouraged to take part in concrete and meaningful interactions with science organisations, industry, universities/academia, government ministries, society – school teachers, pre-college and college students, summer programmes, shared facilities and resources, including long-term maintenance plan and effective use of existing networks

Willingness to Create a different Kind of Institution

By virtue of the location of PAMUST at SIRDC and in Harare, the institution will have ample opportunities for interdisciplinarity and innovation in its research activities as it interacts with various institutions in the city of Harare, the country and indeed the globe through the well-established internet connectivity, among other methods.

Conclusion

The development of African Institutes of Science and Technology (AISTs), in general, is a timely and welcome initiative that should impact positively on the continent's socio-economic development. The establishment of an institute dedicated to the mineral beneficiation and value addition in Southern Africa, in particular, is well conceived and deserves the support of the whole continent and other development partners. Zimbabwe, given its geographical location, the minerals resources at its disposal, the robust support systems to the mining industry, is well positioned and committed to host the Pan-African Minerals Institute of Science and Technology for the benefit of the whole of Africa.

www.ingramcontent.com/pod-product-compliance
Ingram Content Group UK Ltd.
Pitfield, Milton Keynes, MK11 3LW, UK
UKHW022212051026
15624UKWH00001B/131